ENVIRONMENTAL SCIENCE, ENGINEERING AND TECHNOLOGY

NATIONAL FOREST SYSTEM

SPECIAL MANAGEMENT AND ROADLESS AREAS

ENVIRONMENTAL SCIENCE, ENGINEERING AND TECHNOLOGY

Additional books in this series can be found on Nova's website under the Series tab.

Additional E-books in this series can be found on Nova's website under the E-book tab.

ENVIRONMENTAL SCIENCE, ENGINEERING AND TECHNOLOGY

NATIONAL FOREST SYSTEM:

SPECIAL MANAGEMENT AND ROADLESS AREAS

Michael J. Gordon
EDITOR

Nova Science Publishers, Inc.
New York

For permission to use material from this book please contact us:
Telephone 631-231-7269; Fax 631-231-8175
Web Site: http://www.novapublishers.com

Library of Congress Cataloging-in-Publication Data

Alexander, Kristina.
National forest system : special management and roadless areas / editor, Michael J. Gordon.
p. cm.
Authors, Kristina Alexander and Ross W. Gorte.
Originally presented as reports of the Congressional Research Service.
Includes index.
ISBN 978-1-62100-513-1 (hardcover)
1. Forest reserves--United States--Management. 2. Forest roads--Government policy--United States. I. Gordon, Michael J. (Michael Jacob), 1971- II. Gorte, Ross W. III. Library of Congress. Congressional Research Service. IV. Title.
SD426.A63 2011
333.75'110973--dc23
2011035490

Published by Nova Science Publishers, Inc. † New York

CONTENTS

PREFACE

This book explores the National Forest System with a focus on congressionally designated special management areas and roadless area initiatives. Congress authorized the President to reserve public forests to protect the lands and resources. The many presidential proclamations and subsequent land purchases have led to the current National Forest System. These lands are managed to balance the many purposes and values under the Multiple Use-Sustained Yield Act of 1960 through an interdisciplinary planning process, with public involvement, under the National Forest Management Act of 1976. Congress has also designated many specific national forest areas to emphasize particular values or resources, and continues to consider legislation to designate additional specially managed areas within the national forests.

Chapter 1- Congress authorized the President to reserve public forests to protect the lands and resources. The many presidential proclamations and subsequent land purchases have led to the current National Forest System. These lands are managed to balance the many purposes and values under the Multiple Use-Sustained Yield Act of 1960 through an interdisciplinary planning process, with public involvement, under the National Forest Management Act of 1976.

Chapter 2- Roadless areas in the National Forest System (NFS) have received special attention for decades. Many want to protect their relatively pristine condition; others want to use the areas in more developed ways.

Two different roadless area policies have been offered in the last decade. On January 12, 2001, the Clinton Administration's roadless area policy established a nationwide approach to managing roadless areas in the National Forest System to protect their pristine conditions. The Nationwide Rule, as it will be called in this report, generally prohibited road construction and reconstruction and timber harvesting in 58.5 million acres of inventoried roadless areas, with significant exceptions. The Bush Administration initially postponed the effective date of the Nationwide Rule, then issued its own rule that allowed states to plan how roadless areas were managed. It issued a new rule on May 13, 2005. The State Petition Rule allowed governors to petition the Secretary of Agriculture for a special rule for managing the inventoried roadless areas in their states.

In: National Forest System: Special Management...ISBN: 978-1-62100-513-1
Editor: Michael J. Gordon

Chapter 1

CONGRESSIONALLY DESIGNATED SPECIAL MANAGEMENT AREAS IN THE NATIONAL FOREST SYSTEM*

Ross W. Gorte

SUMMARY

Congress authorized the President to reserve public forests to protect the lands and resources. The many presidential proclamations and subsequent land purchases have led to the current National Forest System. These lands are managed to balance the many purposes and values under the Multiple Use-Sustained Yield Act of 1960 through an interdisciplinary planning process, with public involvement, under the National Forest Management Act of 1976.

Congress has also designated many specific national forest areas to emphasize particular values or resources, and continues to consider legislation to designate additional specially managed areas within the national forests. Congress has established two systems of areas with statutory guidelines for all areas. The National Wilderness Preservation System has many units, with general management guidelines that apply to all: no permanent roads or structures; no commercial enterprises; and no motorized or mechanical access. However, the 1964 Wilderness Act that

* This is an edited, reformatted and augmented version of a Congressional Research Service publication, CRS Report for Congress R41285, from www.crs.gov, dated June 11, 2010.

created the Wilderness System and many subsequent wilderness statutes have also included numerous exceptions to these standards for specific areas. The other system is the Wild and Scenic Rivers System, with general management direction for the corridors identified along designated rivers.

In addition, Congress has designated at least 97 individual areas within the National Forest System, and continues to consider more such designations. The Omnibus Public Land Management Act of 2009, for example, designated 10 special areas among its many provisions. While many of the designations are unique, the areas can generally be grouped into six categories: national monuments (6, including one administrative designation); recreation areas (26); scenic areas (11); game refuges (6); protection areas (38); and other (11).

While the statutes designating these 97 areas differ, many provisions are found in multiple statutes. Common administrative provisions include *findings* of the importance and uniqueness of the area; *purposes* for administration of the area; *designation* of the area, often with references to maps and boundaries; *applicable law*, with most making the provisions supplemental to the management guidance for national forests generally; a *management plan* for the area; an *advisory committee* for the area or the plan; authorization of *land acquisition*; and *withdrawal* of the area from the public land laws. Less common administrative provisions include requiring *regulations* for the area; explicitly allowing *inholder access*; prohibiting *buffer zones* around the area; retaining most *state and local governance* provisions; and authorizing *appropriations*.

There are also three common resource management provisions in the many statutes designating special areas: *withdrawal* of the area from mining and mineral leasing laws, subject to valid existing rights; *timber harvesting* restrictions, often with exemptions for fire, insect, or disease control; and *fish and wildlife* management, generally preserving state responsibilities and jurisdiction, allowing fishing and hunting, and authorizing area or period closures for various purposes, in consultation with state officials. Less common resource management provisions include limitations on *motorized access*; authority for *fire, insect, and disease control* actions; continued *livestock grazing*; authority for low-level *military overflights*; and directions on *water use and water rights*.

INTRODUCTION

In 1891, Congress authorized the President to reserve existing public forestlands to protect and preserve the lands and resources. Presidential proclamations of forest reserves (now called national forests), subsequent land purchases, and other congressional authorizations have led to the current National Forest System (NFS)—193 million acres of federal lands in 155 national forests, 20 national grasslands, and numerous other units (land utilization projects, purchase units, research and experimental areas, and more). These lands are generally administered by the Forest Service, in the U.S. Department of Agriculture, for many uses, including timber production, livestock grazing, and many types of recreation. Acceptable levels and locations of various uses are determined with public involvement in a land management planning process.

Congress has also designated specific national forest areas to emphasize particular resources or values. Some types of designations, such as wilderness areas, have statutory guidelines that apply to all designated areas, as well as some site-specific directions. Many other designations, with a variety of titles, are administered under the area-specific guidance enacted when the area was designated.

Congress continues to consider legislation to designate specific national forest areas for particular purposes. For example, the Omnibus Public Lands Management Act of 2009 (P.L. 111-11) designated 20 new national forest wilderness areas, added lands to 20 existing national forest wilderness areas, designated at least nine wild and scenic rivers through national forests, designated two new national scenic areas and one new national recreation area in national forests, and designated seven other areas in national forests with special management provisions.[1]

Additional legislation to designate wilderness and other special areas within the NFS has been introduced.[2] Existing designations and pending legislation differ in the nature of the provisions included and in the level of detail provided. Following a brief overview of NFS management generally, this report summarizes the management provisions for the areas with statutory guidelines for all such designations and then for the areas with statutory guidelines only for each designated area. The Appendix provides a list of all 73 congressionally designated areas with only statutory guidelines for each area, with a summary of each enacted management provision.

MANAGEMENT OF THE NATIONAL FOREST SYSTEM

In 1891, Congress authorized the President to proclaim forest reserves from existing federal lands.[3] Six years later, in what has become known at the Forest Service Organic Administration Act, Congress defined the purposes of the reserves: "No public forest reservation shall be established, except to improve and protect the forest within the reservation, or for the purpose of securing favorable conditions of water flows, and to furnish a continuous supply of timber for the use and necessities of citizens of the United States ..."[4]

Initially the reserves were administered by the Forestry Division of the General Land Office, within the Department of the Interior. In 1905, the 75 million acres of forest reserves (renamed national forests in 1907) [5] were transferred to the Department of Agriculture;[6] the Forestry Division was merged with the existing USDA Bureau of Forestry (whose purpose was to conduct forestry research and to provide assistance to forestland owners) to create a new agency: the U.S. Forest Service (USFS) in the Department of Agriculture. President Theodore Roosevelt increased the acreage of national forests to 172 million acres by 1909. The National Forest System (NFS) has continued to expand slowly, to 193 million acres in 2010.

Multiple-Use Management

The original NFS management goals in the Organic Administration Act were to improve and protect the forest, secure favorable water flows, and furnish timber for citizens. Congress expanded theses goals in the Multiple Use-Sustained Yield Act of 1960 (MUSYA).[7] Section 1 of MUSYA directs that the national forests "shall be administered for outdoor recreation, range, timber, watershed, and wildlife and fish purposes." Section 2 requires that the surface resources be developed and administered for "multiple use and sustained yield," considering the relative values of the various resources, and notes that creating and managing areas of wilderness is consistent with the purposes of the act.

Section 4(a) of MUSYA defines multiple-use management of national forest resources. It directs management to best meet the needs of the American people, with adjustments to changing needs and conditions. Multiple-use management also permits some land to be used for only one or a few resources. Finally, multiple-use management must maintain the productivity of the lands. Thus, MUSYA directs that national forest management balance recreation, timber, grazing, water, and wildlife, as desired by the American people, while sustaining productivity.

Land and Resource Management Planning

Planning for NFS use and management is guided by the National Forest Management Act of 1976 (NFMA).[8] NFMA requires comprehensive land and resource management plans for each unit of the NFS,[9] to be amended as needed and revised periodically but at least every 15 years. The planning requires an interdisciplinary approach combining "physical, biological, economic, and other sciences," directs that the plans provide for multiple use in accordance with MUSYA, and requires public participation in developing, reviewing, and revising the plans.

In Section 6 (16 U.S.C. § 1604), NFMA established numerous requirements and considerations for the planning process. Some provisions were particularly controversial, because of their potential to reduce or otherwise constrain timber harvesting. The more controversial provisions include:

- Section 6(g)(1): "insure that land management plans are prepared in accordance with the National Environmental Policy Act of 1969 ..."[10]
- Section 6(g)(3)(B): "provide for diversity of plant and animal communities based on the suitability and capability of the specific land area ..."
- Section 6(g)(3)(E): "insure that timber will be harvested ... only where—(i) soil, slope, and other watershed conditions will not be irreversibly damaged; (ii) there is assurance that such lands can be adequately restocked within five years after harvest; [and] (iii) protection is provided for streams, streambanks, shorelines, lakes, wetlands, and other bodies of water from detrimental changes in water temperatures, blockages of water courses, and deposits of sediment, where harvests are likely to seriously and adversely affect water conditions or fish habitat ..."
- Section 6(k): "identify lands within the management area which are not suited for timber production, considering physical, economic, and other pertinent factors to the extent feasible, ... [and] assure that, except for salvage sales or sales necessitated to protect other multiple-use values, no timber harvesting shall occur on such lands for a period of 10 years."

Two of these provisions—Section 6(g)(3)(E) and Section 6(k)—were intended to constrain timber harvesting by establishing biophysical and economic requirements, respectively, for timber production. The NEPA provision—Section 6(g)(1)—establishes a procedural requirement that can be, and has been, used to delay or prevent timber harvesting and other development activities in the national forests. The biological diversity provision—Section 6(g)(3)(B)—has perhaps had the most significant impact, however. The 1979 and 1982 regulations to implement this provision required the USFS to maintain viable populations of all native and desirable non-native species. This requirement for viable populations was the initial basis for the litigation that reduced USFS (and BLM) timber harvesting in the Pacific Northwest by 90% or more to maintain the northern spotted owl. (The subsequent listing of the owl under the Endangered Species Act[11] provided further impetus to reduce timber harvesting to preserve northern spotted owl habitat.)

In addition to the controversial provisions of NFMA, efforts to revise the 1982 implementing regulations have been controversial. In November 2000, the Clinton Administration finalized new rules (to be phased in) that emphasized planning for the biological sustainability of the national forests. The Bush Administration delayed implementing the Clinton rules, then replaced them in January 2005 before they went into effect. The final Bush rules were to balance biological and socioeconomic sustainability, to make fewer decisions nationally by reducing regulatory guidelines, and to exempt plans from NEPA and ESA, because the Bush Administration viewed plans as guides to decision-making that would not include site-specific decisions. The Bush planning rules were successfully challenged.[12] In 2007, the USFS reissued the Bush rules as a proposed rule with a draft NEPA environmental impact statement and ESA consultation to meet court requirements. The final planning rules were issued in April 2008, and were invalidated by the U.S. District Court for the Northern District of California in June 2009 for failing to comply with NEPA and ESA.[13] In December 2009, the Obama Administration announced its intent to develop new NFMA planning regulations.[14]

Systems of Special Management Areas

Congress has at times seen fit to alter the general NFS management guidance for particular areas. These areas are commonly designated to emphasize particular resources or values, such as recreation or wildlife, sometimes with detailed congressional guidance for the areas.[15]

As part of its special area designations for some NFS lands, Congress has established two systems of special management areas, with general guidance that applies to all such designated areas: the National Wilderness Preservation System and the National Wild and Scenic Rivers System.[16] These two systems include lands administered by other federal land management agencies, as well as NFS lands, and the general guidance for the designated systems apply to all designated areas.

National Wilderness Preservation System[17]

In 1964, the Wilderness Act[18] created the National Wilderness Preservation System, initially with 54 areas (totaling 9.1 million acres) that the USFS had previously established administratively as wilderness or wild areas.[19] The act also reserved to Congress the right to expand the Wilderness System. Today, the Wilderness System contains 759 wilderness areas totaling nearly 110 million acres in 44 states. Of these, 441 areas with 36.2 million acres (33% of the total) are in the NFS. (See Table 1.) The remaining wilderness areas are administered by the National Park Service, the Fish and Wildlife Service, and the Bureau of Land Management.

The Wilderness Act established general objectives and standards for areas to be included in the Wilderness System. Section 2(c) defines wilderness.

> A wilderness, in contrast with those areas where man and his own works dominate the landscape, is hereby recognized as an area where the earth and its community of life are untrammeled by man, where man himself is a visitor who does not remain. An area of wilderness is further defined to mean in this Act an area of undeveloped Federal land retaining its primeval character and influence, without permanent improvements or human habitation, which is protected and managed so as to preserve its natural conditions and which (1) generally appears to have been affected primarily by the forces of nature, with the imprint of man's work substantially unnoticeable; (2) has outstanding opportunities for solitude or a primitive and unconfined type of recreation; (3) has at least five thousand acres of land or is of sufficient size as to make practicable its preservation and use in an unimpaired condition; and (4) may also contain ecological, geological, or other features of scientific, educational, scenic, or historical value. (16 U.S.C. §1131(c))

Table 1. Wilderness Areas in the National Forest System

State	No.	Acreage	State	No.	Acreage
Alabama	3	41,367	Nevada	21	1,087,624
Alaska	19	5,753,889	New Hampshire	5	138,418
Arizona	36	1,345,008	New Mexico	16	1,387,498
Arkansas	10	116,578	North Carolina	11	102,634
California	64	5,075,290	Oklahoma	2	14,543
Colorado	36	3,140,120	Oregon	41	2,228,103
Florida	7	74,495	Pennsylvania	2	9,002
Georgia	10	114,537	Puerto Rico	1	10,000
Idaho	5	3,961,864	South Carolina	5	16,671
Illinois	7	28,063	South Dakota	1	13,426
Indiana	1	12,463	Tennessee	11	66,349
Kentucky	2	18,132	Texas	5	38,483
Louisiana	1	8,679	Utah	14	775,537
Maine	1	11,232	Vermont	8	100,870
Michigan	10	89,529	Virginia	23	135,325
Minnesota	1	807,853	Washington	25	2,715,056
Mississippi	1	6,046	West Virginia	9	118,810
Missouri	7	64,119	Wisconsin	5	46,414
Montana	12	3,372,525	Wyoming	15	3,111,232
Nebraska	1	7,794	USFS Total	441[a]	36,165,579

Source: USDA Forest Service, *Land Areas of the National Forest System, as of Sept 30, 2009*, Tables 7, 8, and 9, http://www.fs.fed.us/land/staff/lar/2009/lar09index.html.

[a] The total is less than the sum of areas by state, because 13 designated wilderness areas cross state boundaries.

Wilderness Management Guidelines[20]

In addition to the objectives and standards for areas to be designated as wilderness, the Wilderness Act established standards and guidelines for managing designated wilderness areas.[21] In general, each agency administering wilderness areas is to preserve and protect the wilderness character of the areas under its jurisdiction.

To achieve these objectives, Section 4(c) of the Wilderness Act generally prohibits commercial activities; roads, structures, and other infrastructure; and motorized and mechanical access. Specifically, the act states:

> Except as specifically provided for in this Act, and subject to existing private rights, there shall be no commercial enterprise and no permanent road within any wilderness area designated by this Act and, except as necessary to meet minimum requirements for the administration of the area for the purpose of this Act (including measures required in emergencies involving the health and safety of persons within the area), there shall be no temporary road, no use of motor vehicles, motorized equipment or motorboats, no landing of aircraft, no other form of mechanical transport, and no structure or installation within any such area.

Thus, this section prohibits commercial uses, such as timber harvesting. It bars motorized and mechanical access, such as via cars and trucks, bicycles and motorbikes, off-road and all-terrain vehicles, motorboats, and all aircraft. It also prevents construction and maintenance of temporary and permanent roads and structures, such as shelters, water impoundments, fire towers, and communication equipment.

Wilderness Management Exceptions

The Wilderness Act and subsequent statutes designating wilderness areas have included numerous exceptions to the prohibitions described above. In particular, Section 4(d) of the Wilderness Act provides for several specific exceptions:

- Section 4(d)(1) allows continued use of motorboats and aircraft "where these uses have already become established ..."
- Section 4(d)(1) also allows necessary measures for controlling wildfires, and insect and disease infestations, "subject to such conditions as the Secretary deems desirable."
- Section 4(d)(2) allows prospecting for minerals or other resources.
- Section 4(d)(4)(1) allows the President "within a specific area and subject to such regulations as he may deem desirable" to authorize water resource projects and associated infrastructure and access.
- Section 4(d)(4)(2) allows continued livestock grazing, subject to reasonable regulations.

- Section 4(d)(6) allows commercial services “to the extent necessary for activities which are proper for realizing the recreational or other wilderness purposes ...”
- Through December 31, 1983, Section 4(d)(3) allowed establishing mineral rights under mining and mineral leasing laws, subject “to such reasonable regulations as may be prescribed ...” and with patents limited to the minerals (i.e., not to the surface of the land). “Mineral leases, permits, and licenses ... shall contain such reasonable stipulations as may be prescribed ... for the protection of the wilderness character of the land consistent with the use of the land for the purposes for which they were leased, permitted, or licensed.”

Many of the 117 laws designating wilderness areas that have been enacted since 1964 also have exceptions to the no-commerce, no-structures, no-motorized access strictures of the Wilderness Act. Six general types of provisions in many wilderness statutes provide exceptions to the general wilderness management restrictions.[22]

- **Access to private lands.** Section 5 of the Wilderness Act provided for adequate and reasonable access to nonfederal lands and “valid occupancies” (e.g., valid mining claims) within designated wilderness areas. Several subsequent wilderness statutes have essentially reiterated this direction, allowing for adequate access to private lands within designated wilderness areas.
- **Access for wildfire, insect, or disease control.** Section 4(d)(1) of the Wilderness Act explicitly provided for measures needed to control wildfires, insects, and diseases, subject to reasonable regulations. As catastrophic wildfires have apparently become more frequent and more intense in recent years,[23] this authority for nonconforming activities in wilderness areas to control fire, insects, and diseases has essentially been repeated in more recent wilderness statutes.
- **Access for livestock grazing.** Section 4(d)(4)(2) of the Wilderness Act allowed continued livestock grazing in wilderness areas. Continued concerns over access for livestock management led to detailed guidelines for managing livestock grazing in wilderness areas in Appendix A, “Grazing Guidelines,” (pp. 41-43) of H.Rept. 101-405, the Interior and Insular Affairs Committee’s report on the Arizona Desert Wilderness Act of 1990 (P.L. 101-628). The language from the Wilderness Act and from the committee report on the

Arizona Desert Wilderness Act of 1990 has subsequently been repeated or referenced in several statutes.

- **Native and tribal rights and access.** Several wilderness statutes have provided for access to the designated wilderness areas for native traditional cultural and religious purposes. The guidance has either been that nothing in the act "alters, modifies, enlarges, diminishes, or abrogates the treaty rights ..." or that the Secretary is to "ensure access for traditional cultural and religious purposes" and to allow temporary closures (i.e., exclusive access) for the smallest possible area and minimum possible period. Provisions in other statutes have also directed cooperation, memoranda of understanding, and agreements to provide access to and protection of important tribal use areas.
- **Low-level military overflights.** Since 1990, several wilderness statutes have included provisions to specify that nothing in the act "shall preclude low level overflights of military aircraft, the designation of new units of special airspace, or the use or establishment of military flight training routes over wilderness areas designated by this" act.
- **Access for hydrological, meteorological, and climatological equipment.** Several wilderness statutes have included provisions allowing the installation and maintenance of weather-related equipment and facilities within one or more of the areas designated by the statute.
- **Other provisions.** Many statutes have included specific exceptions, usually for a specific area, commonly to allow motorized access for wildlife management or for maintaining water resource infrastructure; to allow continued motorboat access; to allow aircraft for specific purposes; or to allow establishment and maintenance of water infrastructure or telecommunication equipment.

National Wild and Scenic Rivers System

The National Wild and Scenic Rivers System was established in the Wild and Scenic Rivers Act of 1968.[24] The act established a policy of preserving designated free-flowing rivers and their immediate environments "for the benefit and enjoyment of present and future generations." Rivers are generally added to the system by an act of Congress, although states can nominate, and the Secretary of the Interior can approve, rivers protected under state law and managed by a state agency. The Rivers System includes three classes of rivers—wild, scenic, and recreational— depending on the extent and nature of development along the river segment.

The administering agency establishes boundaries for the designated river's corridor, averaging not more than a half-mile in width (including the river and adjacent shoreline). Management of the lands within the corridor depends, in part, on the river's classification; developmental activities are most restricted along wild rivers and least restricted along recreational rivers. Administration is to emphasize the values that led to the designation, and uses are generally restricted only if they harm those values. Section 10(a) of the Wild and Scenic Rivers Act states:

> Each component of the national wild and scenic rivers system shall be administered in such a manner as to protect and enhance the values which caused it to be included in said system without, insofar as is consistent therewith, limiting other uses that do not substantially interfere with public use and enjoyment of these values. In such administration primary emphasis shall be given to protecting its esthetic, scenic, historic, archaeologic, and scientific features.

Thus, the USFS is to manage the national forest lands within the National Wild and Scenic Rivers System under MUSYA and NFMA, as long as the activities are consistent with protecting the wild, scenic, or recreational nature of the designated rivers. Table 2 shows the area of wild and scenic river corridors within the National Forest System in each state with one or more designated rivers.

Table 2.Wild and Scenic River Corridors in the National Forest System

State	No.[a]	Acreage	State	No.[a]	Acreage
Alabama	1	8,780	Missouri	1	8,978
Arizona	2	16,937	Montana	1	40,627
Arkansas	8	34,364	New Hampshire	1	1,049
California	21	309,956	New Mexico	4	16,133
Colorado	1	15,141	North Carolina	3	5,616
Georgia	1	6,786	Oregon	45	326,856
Idaho	6	174,596	Pennsylvania	2	8,221

State	No.[a]	Acreage	State	No.[a]	Acreage
Kentucky	1	3,880	Puerto Rico	3	2,848
Louisiana	1	5,150	South Carolina	1	7,041
Michigan	15	125,410	Washington	3	28,282
Mississippi	1	4,599	Wyoming	2	107,481
			USFS Total[b]	124	1,258,731

Source: USDA Forest Service, *Land Areas of the National Forest System, as of Sept 30, 2009*, Tables 7, 8, and 9, http://www.fs.fed.us/land/staff/lar/2009/lar09index.html.

[a] Includes the number of rivers regardless of designation; for example, the Verde River (AZ) has a segment designated as a wild river and another segment designated as a scenic river, but this is counted as one river in this report.

[b] The total is less than the sum of rivers by state, because two designated wild and scenic rivers occur in more than one state.

Individual Special Management Areas

In addition to the two standardized systems of special management areas, Congress has designated numerous individual special management areas (SMAs) with management guidelines for each area. Six major types of designations have been enacted: national monuments; recreation areas; scenic areas; game refuges;[25] protection areas; and other. These areas differ from the two standardized systems (wilderness and wild and scenic river corridors) in having neither general authorizing legislation for the type of designation nor general management direction for all such areas. Rather, each SMA has been created by an act of Congress with its own management guidelines and restrictions. Furthermore, there are no obvious differences in the guidelines and restrictions for the different types of designations. Table 3 shows the number of areas for each of the five major types of SMAs, and the total area designated. The Appendix provides a complete list of the 98 SMAs designated to date, with a quotation or summary of the major provisions enacted for each.

Table 3. Special Management Areas in the National Forest System

Designation	No.	Acreagea
National monuments	6	2,827,634
Recreation areas	26	3,199,000
Scenic areas	11	324,853b
Game refuges	6	884,452c
Protection areas	38	1,059,716d
Other designations	11	277,451e
Total	98	9,573,106

Source: Appendix and USDA Forest Service, *Land Areas of the National Forest System, as of Sept 30, 2009*, Tables 11, 12, 15-20, and 22-26, http://www.fs.fed.us/land/staff/lar/2009/lar09index.html.

[a] Acreage as reported in *Land Areas of the National Forest System*, where available.

[b] Excludes two areas for which acreage is not reported in statute or in *Land Areas of the National Forest System.*

[c] Excludes six areas with 240,932 acres proclaimed by the President.

[d] Excludes seven areas for which acreage is not reported in statute or in *Land Areas of the National Forest System.*

[e] Excludes four areas for which acreage is not reported in statute or in *Land Areas of the National Forest System.*

Numerous provisions have been used commonly in many of the statutes designating SMAs. The common provisions can be grouped into administrative provisions and resource management provisions. In addition, some provisions have been used in a few of the SMA statutes, and can similarly be grouped into administrative and resource management provisions.

Common Administrative Provisions

Eight types of administrative provisions have commonly been enacted for SMAs in the National Forest System. The details of the language used often differs, especially for the first two types of provisions, but the intent and effect is usually quite similar.

1) **Findings.** Many SMA statutes include findings about the unique attributes of the area being designated.

2) **Purpose.** Most SMA statutes include a statement about the purpose(s) for designating the area. Such statements often provide a basis for management that may differ from management of the area prior to designation.
3) **Designation.** Many SMA statutes include a provision explicitly designating and naming the area. These provisions often include a reference to the area's boundaries and to a map (commonly identified by a name and date) to be retained in a specific office (typically the Forest Supervisor's office for the national forest in which the SMA is located).
4) **NFS Laws.** Most SMA statutes direct that the SMA is to be managed consistent with the laws, rules, and regulations generally applicable to National Forest System lands. Thus, the purposes and provisions enacted for the designated area are generally supplemental to existing governance, and not a substitute for that management guidance.
5) **Management Plan.** Many SMA statutes require a management plan for the designated area, commonly completed within a specified period. Sometimes the planning guidance requires public participation; sometimes it does so indirectly, by reference to the NFMA planning process and requirements.
6) **Advisory Committee.** Many SMA statutes provide for an advisory committee or group for the designated area. This advisory committee is occasionally permanent, but sometimes is just for the development of the management plan. Also, some advisory groups are to represent the array of general public interests, but others are to focus on scientific issues associated with management of the area.
7) **Land Acquisition.** Most SMA statutes authorize land acquisition for nonfederal land within the boundaries of the area. As the USFS has general land acquisition authority under the Weeks Law (16 U.S.C. § 515 et al.) and the SMAs are within the NFS, such land acquisition authority might be redundant.
8) **Withdrawal.** Many of the SMA statutes withdraw the lands of the designated area from access under public land laws, such as the Homestead Act. As most of the public land laws were repealed in the Federal Land Policy and Management Act of 1976 (FLPMA)[26] and are no longer applicable to NFS or other federal lands, the more recent SMA statutes are less likely to include such a provision.

Common Resource Management Provisions

Three resource provisions have been included in many of the statutes designating SMAs in the NFS. These provisions generally address management of, and access to, minerals, timber, and fish and wildlife.

1) **Minerals.** Most SMA statutes withdraw the designated lands from access under the mining and the mineral and geothermal leasing laws, subject to valid existing rights. This is very similar to the provisions of the Wilderness Act for withdrawing lands from access for mineral exploration and development. Some statutes authorize the USFS to continue to sell mineral materials (sand and gravel) from within the designated area.
2) **Timber.** Many SMA statutes restrict timber harvesting. The most common approach is to use the restrictions of the Wilderness Act: no commercial timber harvesting, except as needed for fire, insect, and disease control and for visitor and administrative safety. Other approaches include requiring *uneven-aged* management practices,[27] directing that harvesting be adjusted for possible scenic impacts, or limiting the size or species of trees that can be cut and removed.
3) **Fish and Wildlife.** Most of the statutes designating SMAs include provisions relating to fish and wildlife management. In general, the statutes explicitly retain state jurisdiction over the management of the animals, limiting federal SMA administration to habitat management. Many of the statutes also explicitly allow hunting and fishing, and sometimes trapping, consistent with state law. These provisions also typically allow seasonal and/or area closures to protect critical wildlife needs (e.g., elk calving grounds), usually in consultation with the relevant state agency.

Less Common Administrative Provisions

Several additional administrative provisions have been enacted in at least a few statutes designating SMAs.

1) **Regulations.** Several SMA statutes direct that the Secretary of Agriculture, acting through the Chief of the Forest Service, promulgate regulations for administering the designated area.

2) **Inholder Access.** Some SMA statutes explicitly allow access for nonfederal landowners whose lands are within the designated SMA. In some instances, the language parallels the Wilderness Act, allowing "reasonable" rights of access.
3) **Buffer Zones.** Some of the statutes designating SMAs prohibit the USFS from establishing buffer zones around the SMA. This provision is akin to provisions enacted in several statutes designating wilderness areas, which note that the ability to see or hear a nonconforming activity from within the wilderness is not sufficient authority to restrict or prohibit that activity.
4) **Effect of Governance.** Several SMA statutes direct that the designation is to have no effect on state or local laws and regulations, including law enforcement and taxation of private enterprise within the SMA.
5) **Authorization of Appropriations.** Some SMA statutes explicitly authorize appropriations for administering the area consistent with the statute. However, since few, if any, SMA statutes authorize any activity not already authorized for NFS lands generally (typically they restrict already-authorized activities), and since USFS appropriations are enacted for activities, not for specific units, it is not clear that an authorization is required for an SMA.

Less Common Resource Management Provisions

In addition to administrative provisions, several of the statutes designating SMAs have included additional provisions governing natural resource management within the SMA.

1) **Motorized Access.** Several SMA statutes prohibit motorized access in the area, except on existing roads and other routes designated for motorized access. In addition, some of the statutes also prohibit the construction of new roads within the SMA. Exceptions to this prohibition have been included for new roads to replace existing, environmentally damaging roads and for new roads as decided in the planning process for the area.
2) **Wildfire, Insect, and Disease Control.** Some SMA statutes explicitly allow activities to control fires, insects, and diseases. In general, these provisions use the same language as the Wilderness

Act—necessary measures subject to reasonable restrictions as determined by the Secretary.

3) **Livestock Grazing.** Some of the statutes designating SMAs explicitly allow livestock grazing to continue where it existed prior to the designation. Motorized access, even in areas where such access is generally prohibited, is generally permitted, consistent with historic use. The provisions, when enacted, are commonly similar in intent to the provisions in the Wilderness Act.
4) **Military Overflights.** Some SMA statutes explicitly direct that the designation is not to prohibit or otherwise affect low-level military training overflights of the SMA. Again, the language is generally the same as is used in the several statutes designating wilderness areas.
5) **Water Use/Water Rights.** A few statutes designating SMAs specifically address water use and water rights associated with the area. These provisions are more difficult to summarize, as they are more variable than many of the other provisions, ranging from no effect on water rights and water use, to an explicit reservation of water as needed for the SMA, to an explicit denial of water rights for the SMA.

APPENDIX. INDIVIDUAL CONGRESSIONALLY DESIGNATED AREAS IN THE NATIONAL FOREST SYSTEM

This appendix lists the 98 congressionally designated areas, grouped into six major types of designations: national monuments; recreation areas; scenic areas; game refuges; protection areas; and other. The areas are listed chronologically for each type of designation. For each area, the entry displays the statute designating the area and the length of its statutory provisions; the length cited for each is the number of pages in the *U.S. Code Congressional and Administrative News*, by West Publishing Co. (St. Paul, MN), or in the free-standing slip law produced by the U.S. Government Printing Office, except as otherwise noted.[28] Each entry then identifies the management purposes, provisions, requirements, and prohibitions with quotations or brief CRS annotations.

National Monuments

Four national monuments and two national volcanic monuments have been established within the National Forest System.[29] Five have been designated by acts of Congress; one was established by presidential proclamation under the Antiquities Act of 1906 (16 U.S.C. §§ 431 et seq.).[30] No conclusions can be drawn about the general management guidance for these areas, or the differences compared to recreation or scenic areas. The sizes of the national monuments vary widely, from about 55,000 acres for the Newberry National Volcanic Monument to nearly 2.3 million for the Misty Fjords National Monument.

1 & 2. Misty Fjords National Monument and Admiralty Island National Monument (AK)—P.L. 96-487, Alaska National Interests Lands Conservation Act of 1980 (ANILCA); 13 pages.

- §503(a)—designates Misty Fjords, 2,285,000 acres.
- §503(b)—designates Admiralty Island, 921,000 acres.
- §503(c)—directs management as "units of the National Forest System to protect objects of ecological, cultural, geological, historical, prehistorical, and scientific interest," subject to valid existing rights.
- §503(d)—prohibits "the sale or harvesting of timber: *Provided*, That nothing ... shall prevent ... measures as may be necessary for the control of fire, insects, and disease." (emphasis in original)
- §503(f)(1)—withdraws areas from mining, mineral leasing, and public land laws, subject to valid existing rights; (2) valid existing rights are subject to "reasonable regulations ... to assure that such activities are compatible, to the maximum extent feasible, with the purposes."
- §503(h) & (i)—provides lengthy [2+ pages] direction and requires environmental impact statement for surface access road for Quartz Hill mineral deposit.
- §503(j)—directs a special use permit for Thayer Lake Lodge.
- §504—guides treatment of unperfected mining claims in the monuments [21/2 pages].
- §505(a)—directs that fishing is subject to "reasonable regulations ... to maintain the habitats, to the maximum extent feasible, of anadromous fish and other foodfish, and to maintain the present and continued productivity of such habitat when such habitats are affected by mining activities."

- §505(b)—has no effect on state responsibility and authority for fish and wildlife.
- §506—directs land exchanges [5+ pages].
- §507—requires cooperative fisheries planning.

3. Mount St. Helens National Volcanic Monument (WA)—P.L. 97-243; 8+ pages.

- §1—designates 110,000-acre area.
- §3—directs land acquisition [31/2 pages].
- §4(a)—directs management "as a separate unit ... in accordance with the appropriate laws pertaining to" the NFS.
- §4(b)—establishes purposes: (1) "to protect geologic, ecologic, and cultural resources" and (2) "control fire, insects, diseases and other agents that might (A) endanger irreplaceable features ... or (B) cause substantial damage to significant resources", and measures "reasonably necessary to ensure public safety and prevent loss of life and property."
- §4(c)—authorizes "full use ... for scientific study and research."
- §4(d)—allows control of "times and means of access and use" but does not prohibit "the use of motorized vehicles, aircraft or motorboats for emergency and other essential administrative services, including those provided by State and local governments, or when necessary, for authorized research."
- §4(e)(1)—requires "recreational use ... [and] recreational and interpretive facilities (including trails and campgrounds)."
- §4(e)(2)—directs that "roads and other facilities ... should be located generally in areas which were developed prior to the 1980 eruption."
- §4(f)—withdraws area from mining, mineral leasing, and public land laws, subject to valid existing rights.
- §4(g)—allows no timber harvesting, except existing salvage contracts and the "minimum extent necessary to control fire, insects, diseases, and other agents."
- §4(h)—permits hunting and fishing. Allows zones and periods of no hunting or fishing for public safety, resource protection, scientific research, or public use and enjoyment; requires consultation with the state. Has no effect on state jurisdiction or responsibility for fish and wildlife.

- §4(i)—requires a "detailed and comprehensive management plan" within three years, "integrated and periodically revised" with the NFMA plan for the Gifford Pinchot National Forest.
- §6—creates no buffer zone.
- §7—establishes a scientific advisory board; defines purpose and membership.
- §8—directs county payment allocation to roads and schools.
- §9—authorizes appropriations.

4. Newberry National Volcanic Monument (OR)—P.L. 101-522; 8+ pages.

- §1(a)—establishes purposes: "to preserve and protect for present and future generations its remarkable geologic landforms and for ... providing for the conservation, protection, interpretation, and enhancement of its ecological, botanical, scientific, scenic, recreational, cultural, and fish and wildlife resources."
- §1(b)—establishes five designations: Monument; Transferal Area; Transferal Corridor; Transferal Area Adjacent;[31] and Special Management Area.
- §2(a)—directs management "in accordance with the laws, rules, and regulations applicable to the National Forest System."
- §2(b)—identifies areas added to the monument upon "termination, cancellation, or relinquishment" of geothermal leases.
- §2(e)—identifies management requirements:
  - "(1) shall allow natural ecological succession.... Timber removal shall be permitted only ... [as] necessary to achieve the purposes of this Act and to protect health and safety."
  - "(2) Recreation uses and interpretive facilities shall be provided."
  - "(3) Roads shall be permitted ... consistent with the purposes of this Act and in accordance with the management plan."
- §2(f)—directs that "scientific research shall be allowed."
- §2(g)—directs "action to the extent practicable to ensure that tree diseases, insect infestations, fire hazards, and fires ... do not seriously threaten resources outside the Monument ..."
- §3—directs exchanges of geothermal lease rights.
- §4—withdraws areas from mineral leasing, mining, and public land laws, subject to valid existing rights.

- §5—has no effect on state fish and wildlife jurisdiction or responsibilities, including regulation of hunting, fishing, and trapping. Allows zones and periods of no hunting for public safety, administration, or public use and enjoyment; requires consultation with state.
- §6(a)—requires a management plan within 3 fiscal years, in consultation with federal, state, and local agencies, and the public. The plan is to be updated, amended, or revised as needed, "separately or in conjunction with" the NFMA plan for the Deschutes National Forest.
- §6(b)—requires that management planning consider:
    - "(1) Recreation, including consideration of a full range of existing and appropriate new facilities ..."
    - "(2) Vegetation, including consideration of a full range of management options, and a program to reestablish old growth ponderosa pine ecosystems."
    - "(3) Roads and facilities, including ... (D) criteria for the closing and obliteration of roads."
    - "(4) Fire and fuel management prescriptions."
    - "(5) Wildlife management, including general prescriptions for wildlife habitat improvements."
    - "(6) Research."
    - "(7) Monitoring, including monitoring needs for air, water, wildlife, soil, and other resources."
    - "(8) Conflicts ... among uses and resources."
- §7—establishes an advisory council for advice on initial management plan; identifies membership.
- §8—directs no buffer zone and no effect on existing contracts.
- §9—provides definitions.
- §10—authorizes appropriations.

5. Giant Sequoia National Monument (CA)—Proclamation 7295 of April 15, 2000, under the authority of the Antiquities Act of 1906 (16 U.S.C. §431 et seq.); 4+ pages in 65 Fed. Reg. 24095, 4/25/2000).

- 1—proclaims 327,769 acres for "the purpose of protecting the objects identified in the above preceding paragraphs ..." [10 unnumbered paragraphs providing findings].

- 2—withdraws area from public land, mining, and mineral leasing laws.
- 3—makes management subject to valid existing rights.
- 4—directs completion timber contracts signed prior to January 1, 1999. States that the monument is not "suited for timber production.... Removal of trees, except for personal use fuel wood, ... may take place only if clearly needed for ecological restoration and maintenance or public safety."
- 5—directs management "pursuant to applicable legal authorities." Requires a management plan within three years and regulations to "provide for and encourage continued public and recreational access and use consistent with the purposes."
- 6—creates a "Scientific Advisory Board to provide scientific guidance during the development of the initial management plan." Requires consultation with National Academy of Sciences in forming the board.
- 7—requires the management plan to including a transportation plan "for visitor enjoyment and understanding about the scientific and historic objects in the monument, consistent with their protection." Motorized use allowed only on designated roads and trails, except for emergency and authorized administrative purposes. Bans new roads or trails.
- 8—has no effect on fish and wildlife jurisdiction.
- 9—reserves "a quantity of water sufficient to fulfill the purposes," subject to valid existing rights.
- 11—has no effect on existing special use authorizations.
- 13—includes "Warning ... not to appropriate, injure, destroy, or remove any feature ... and not to locate or settle upon any of the lands."

6. Santa Rosa and San Jacinto Mountains National Monument (CA)—P.L. 106-351; 7+ pages. Jointly managed with BLM.

- §2(a)—presents findings.
- §2(b)—designates area "to preserve the nationally significant biological, cultural, recreational, geological, educational, and scientific values ... and to secure now and for future generations the opportunity to experience and enjoy the magnificent vistas, wildlife,

land forms, and natural and cultural resources in these mountains and to recreate therein."

- §3—(a) directs management in accordance NFMA and "other applicable provisions of law;" (c) has no effect on Indian reservation; (d) is subject to valid existing rights; (e) has no buffer zones; and (f) has no effect on air and water quality standards.
- §4—requires a management plan within three years, with consultation and cooperation.
- §5(a)—directs actions to "continue to authorize ... such recreational uses as hiking, camping, mountain biking, sightseeing, and horseback riding."
- §5(b)—allows motorized use only on designated roads and trails, except for administration or emergency use.
- §5(c)—permits hunting, trapping, and fishing. Allows zones and periods of no hunting, trapping, or fishing for public safety, administration, or public use and enjoyment.
- §5(d)—requires "adequate access to nonfederally owned lands or interests in land within the boundaries."
- §5(e)—has no effect on valid existing public utility rights-of-way.
- §5(f)—requires the management plan to "address the maintenance of roadways, jeep trails, and paths."
- §5(g)—has no effect on grazing use.
- §5(h)—has no effect on overflights, but permits no new commercial air tour operations.
- §5(i)—withdraws area from public land, mining, and mineral leasing and mineral materials laws, subject to valid existing rights.
- §6—authorizes land acquisition.
- §7—establishes a local advisory committee for preparing and implementing the management plan.
- §8—authorizes appropriations.

Recreation Areas

Congress has designated 26 recreational areas within the National Forest System, each with distinct management direction enacted in the law establishing the area. Most have been named *national recreation areas*, although two are *recreation areas*, one is a *recreational area*, one is a *scenic recreation area*, and one is a *winter recreation area*. The rationale for these distinctions is not apparent. Also, no conclusions can be drawn about the general management guidance for these areas, or the differences compared to national monuments or scenic areas. The sizes of the recreational areas vary widely, from less than 10,000 acres for three units to more than 700,000 acres for the Sawtooth NRA.

1. Spruce Knob-Seneca Rocks National Recreation Area (WV)—P.L. 89-207; 2 pages.

- §2—designates an area "not to exceed 100,000 acres."
- §3—authorizes land acquisition.
- §4—directs "an accelerated program of development of facilities for outdoor recreation."
- §5—directs management "in accordance with the laws, rules, and regulations applicable to the national forests ... [as] will best provide for (1) public outdoor recreation benefits; (2) conservation of scenic, scientific, historic, and other values contributing to public enjoyment; and such management, utilization, and disposal of natural resources as ... will promote, or is compatible with, and does not significantly impair the purposes."
- §6—permits hunting and fishing. Allows zones and periods of no hunting for public safety, administration, or public use and enjoyment; requires consultation with state.

2. Whiskeytown-Shasta-Trinity National Recreation Area (CA)—P.L. 89-336; 6+ pages. Jointly managed with National Park Service.

- §1—designates area "to provide ... for the public outdoor recreation use and enjoyment ... by present and future generations and the conservation of scenic, scientific, historic, and other values contributing to public enjoyment." Requires coordinated planning and administration "to provide integrated management policies for the recreation area as a whole."
- §2—authorizes acquisition of lands and easements; prohibits condemnation of "improved property."
- §4(a)—requires management coordinated with the Central Valley Project and for area's purposes. Directs "administration, protection, and development ... [as] will best provide for (1) public outdoor recreation benefits; (2) conservation of scenic, scientific, historic, and

other values contributing to public enjoyment; and such management, utilization, and disposal of natural resources as ... will promote, or is compatible with, and does not significantly impair the purposes."

- §4(b)(1)—requires use of "statutory authorities relating to the national forests in such a manner as ... appropriate to carry out the purposes."
- §5—permits hunting and fishing. Allows zones and periods of no hunting for public safety, administration, or public use and enjoyment; requires consultation with state.
- §6—withdraws area from mining, subject to valid existing rights; "may permit the removal of the nonleasable minerals" in accordance with §10 of Act of September 1, 1939 (43 U.S.C. §192c) and "of leasable minerals ... in accordance with the Mineral Leasing Act of February 25, 1920 (30 U.S.C. §181 et seq.), if he [the Secretary] finds that such disposition would not have significant adverse effects on the purposes of the Central Valley Project or the administration of the recreation area."
- §7—has no effect on state or local "right to exercise civil and criminal jurisdictions within the recreation area or of its right to tax persons, corporations, franchises, or property."
- §10—authorizes appropriations.

3. Mount Rogers National Recreation Area (VA)—P.L. 89-438; 11/2 pages.

- §1—establishes purposes: "to provide ... for the public outdoor recreation use and enjoyment ... by present and future generations and the conservation of scenic, scientific, historic, and other values contributing to public enjoyment."
- §2—designates area.
- §3—directs acquisition of "lands, waters, or interests therein, including scenic and other easements within the boundaries."
- §4—directs "an accelerated program of development of facilities for outdoor recreation."
- §5—directs "administration, protection, and development ... [as] will best provide for (1) public outdoor recreation benefits; (2) conservation of scenic, scientific, historic, and other values contributing to public enjoyment; and such management, utilization, and disposal of natural resources as ... will promote, or is compatible with, and does not significantly impair the purposes."

- §6—permits hunting and fishing. Allows zones and periods of no hunting for public safety, administration, or public use and enjoyment; requires consultation with state.

4. Flaming Gorge National Recreation Area (UT-WY)—P.L. 90-540; 21/2 pages.

- §1—designates area "to provide ... for the public outdoor recreation use and enjoyment ... and the conservation of scenic, scientific, historic, and other values contributing to public enjoyment."
- §2—directs "administration, protection, and development ... in accordance with the laws, rules, and regulations applicable to national forests, in a manner coordinated with the other purposes of the Colorado River storage project ... [as] will best provide for (1) public outdoor recreation benefits; (2) conservation of scenic, scientific, historic, and other values contributing to public enjoyment; and (3) such management, utilization, and disposal of natural resources as ... will promote, or is compatible with, and do not significantly impair the purposes."
- §4—permits hunting, fishing, or trapping. Allows zones and periods of no hunting, fishing, or trapping for public safety, administration, or public use and enjoyment; requires consultation with state. Has no effect on state jurisdiction or responsibilities for hunting and fishing.
- §5—withdraws area from mining, subject to valid existing rights; "may permit the removal of the nonleasable minerals ... in the manner prescribed by" §10 of the Act of September 1, 1939, as amended, (43 U.S.C. §387) and "of leasable minerals ... in accordance with the Mineral Leasing Act of February 25, 1920, as amended (30 U.S.C. §181 et seq.), if he [the Secretary] finds that such disposition would not have significant adverse effects on the purposes of the Colorado River storage project and ... the purposes of the recreation area."
- §9—has no effect on state or local "right to exercise civil and criminal jurisdiction within the recreation area ... or of its right to tax persons, corporations, franchises, or other non-Federal property."

5. Oregon Dunes National Recreation Area (OR)—P.L. 92-259; 41/2 pages.

- §1—designates area "to provide for the public outdoor recreation use and enjoyment of certain ocean shorelines and dunes, forested areas, fresh water lakes, and recreational facilities ... by present and future generations and the conservation of scenic, scientific, historic, and other values contributing to public enjoyment."
- §2—requires management "in accordance with the laws, rules, and regulations applicable to the national forests."
- §3—establishes the "Inland Sector" as a buffer zone.
- §7—authorizes land acquisition of certain types of land in the Inland Sector only with the consent of the owner.
- §9—permits hunting, fishing, and trapping. Allows zones and periods of no hunting, fishing and trapping for public safety, administration, or public use and enjoyment; requires consultation with state.
- §10—withdraws area from mining and mineral leasing laws, subject to valid existing rights.
- §11(a)—authorizes and directs "subject to applicable water quality standards ... [and] to reasonable rules and regulations, the investigation for, appropriation, storage, and withdrawal of ground water, surface water, and lake, stream, and river water ... for beneficial use in accordance with applicable laws of the United States and of the State of Oregon."
- §11(b)—authorizes and directs "subject to applicable water quality standards ...transportation and storage in pipelines ... of domestic and industrial wastes."
- §11(c)—authorizes "subject to applicable water quality standards ... such additional easements and rights ... for the effective use of the rights to water and the disposal of waste ... and for other utility and private purposes" with approval from the state "subject to such reasonable terms and conditions as he [the Secretary] deems necessary for the protection of the scenic, scientific, historic, and recreational features."
- §12—establishes advisory council for management and development.
- §13—directs review of wilderness potential within three years.
- §14—requires cooperation with state and local governments. Also, has no effect on state or local "right to exercise civil and criminal jurisdictions within the recreation area ... or of its right to tax persons, corporations, franchises, or other non-Federal property."
- §15—authorizes appropriations.

6. Sawtooth National Recreation Area (ID)—P.L. 92-400; 5+ pages.

- §1—designates area "to assure the preservation and protection of the natural, scenic, historic, pastoral, and fish and wildlife values and to provide for the enhancement of the recreational values associated therewith."
- §2(a)—directs management "in accordance with the laws, rules and regulations applicable to the national forests ... as will best provide (1) the protection and conservation of the salmon and other fisheries; (2) the conservation and development of scenic, natural, historic, pastoral, wildlife, and other values, contributing to and available for public recreation and enjoyment, including the preservation of sites associated. with and typifying the economic and social history of the American West; and (3) the management, utilization, and disposal of natural resources ... such as timber, grazing, and mineral resources insofar as their utilization will not substantially impair the purposes."
- §2(b)—designates wilderness.
- §3—authorizes land acquisition, up to 5% of private property within the recreation area; (c) limits condemnation.
- §4—requires "regulations setting standards for the use, subdivision, and development of privately owned property within the boundaries ... in furtherance of the purposes" in conformance with the Administrative Procedure Act, and with any litigation at the U.S. District Court for Idaho.
- §5—requires review of wilderness suitability.
- §7—has no effect on state or local right "to exercise civil and criminal jurisdictions within the recreation area or of its right to tax persons, corporations, franchises, or property."
- §8—permits hunting and fishing. Allows zones and periods of no hunting for public safety, administration, or public use and enjoyment; requires consultation with state.
- §9—creates "no express or implied claim or denial on the part of the Federal Government as to exemption from State water laws."
- §10—withdraws area from mining laws, subject to valid existing rights.
- §11—requires regulations including "provisions for control of the use of motorized and mechanical equipment for transportation over, or alteration of, the surface of such federal land ..., including but not

limited to mineral prospecting, exploration, or development operations."

- §12—prohibits minerals land patents.
- §13—authorizes appropriations.
- §14—directs a "comprehensive analysis of the natural, economic, and cultural values of the recreation area and adjacent Pioneer Mountains for ... evaluating the potentiality" as a National Park Service unit.

7. Hells Canyon National Recreation Area (ID-OR)—P.L. 94-199; 6+ pages.

- §1(a)—designates area "to assure that the natural beauty, and historical and archeological values ... are preserved for this and future generations, and that the recreational and ecological values and public enjoyment of the are thereby enhanced."
- §2 & §3—designate wilderness areas and wild and scenic rivers, respectively.
- §4—directs that "the Federal Power Commission[32] may not license the construction of any dam, water conduit, powerhouse, transmission line, or other project work ... within the area" but operation and maintenance of current projects may continue.
- §6—creates no limits or restrictions on beneficial uses of upstream water, and no flow requirements.
- §7—directs management "in accordance with the laws, rules, and regulations applicable to the national forests, for public outdoor recreation in a manner compatible with the following objectives:"
  - "(1) the maintenance and protection of the free-flowing nature of the rivers;"
  - "(2) conservation of the scenic, wilderness, cultural, scientific, and other values contributing to the public benefit;"
  - "(3) preservation ... of all features and peculiarities believed to be biologically unique including, but not limited to, rare and endemic plant species, rare combinations of aquatic, terrestrial, and atmospheric habitats, and the rare combinations of outstanding and diverse ecosystems and parts of ecosystems associated therewith;"
  - "(4) protection and maintenance of fish and wildlife habitat;"

- "(5) protection of archeological and paleontological sites and interpretation of these sites for the public benefit and knowledge insofar as it is compatible with protection;"
- "(6) preservation and restoration of historic sites associated with and typifying the economic and social history of the region and the American West; and"
- "(7) such management, utilization, and disposal of natural resources ... including, but not limited to, timber harvesting by selective cutting, mining, and grazing and the continuation of such existing uses and developments as are compatible with the provisions of this Act."

- §8(a)—requires comprehensive management plan within five years.
  - (b)—requires consideration of "the historic, archeological, and paleontological resources ... which offer significant opportunities for anthropological research." Requires an inventory.
  - (c)—directs a "detailed study of the need for, and alternative routes of, scenic roads and other means of transit to and within the recreation area."
  - (d)—requires a review of wilderness suitability.
  - (e)—requires "full public participation."
  - (f)—directs that "such activities as are compatible with the provisions of this Act, but not limited to, timber harvesting by selective cutting, mining, and grazing may continue during development of the comprehensive management plan."
- §9—authorizes acquisition of land or easements; authorizes purchase land without consent if reasonable negotiations have failed and up to 5% of private land; places no such limits on easements, but authorizes easements without consent only if "such lands are being used, or are in imminent danger of being used, in a manner incompatible" with the regulations.
- §10—requires "such rules and regulations as he deems necessary ... including, but not limited to—"
  - "(a) standards for the use and development of privately owned property within the recreation area;"
  - "(b) standards and guidelines to insure full protection and preservation of the historic, archeological, and paleontological resources in the recreation area;"
  - "(c) provision for the control of the use of motorized and mechanical equipment;"

    - "(d) provision for the control of the use and number of motorized and nonmotorized river craft; and"
    - "(e) standards for such management, utilization and disposal." See §8(f).
- §11—withdraws area from mining and mineral leasing laws, subject to valid existing rights.
- §12—permits hunting and fishing. Allows zones and periods with no hunting or fishing for public safety, administration, or public use and enjoyment; requires consultation with state.
- §13—recognizes "ranching, grazing, farming, timber harvesting, and the occupation of homes and lands associated therewith ... as traditional and valid uses."
- §14—has no effect on state or local rights "to exercise civil and criminal jurisdictions within the recreation area or of its right to tax persons, corporations, franchises, or property."
- §15—requires cooperation "in the development and operation of facilities and services."
- §16—authorizes appropriations.

8. Arapaho National Recreation Area (CO)—P.L. 95-450, Indian Peaks Wilderness Area, the Arapaho National Recreation Area and the Oregon Islands Wilderness Area Act; 21/2 pages.

- §2—(1) presents findings and establishes purposes: "to preserve and protect the natural, scenic, historic, pastoral, and wildlife resources of the area and to enhance the recreational opportunities provided."
- §3—designates wilderness.
- §4(a)—designates 36,235-acre area.
- §4(b)—requires management "in accordance with the laws and regulations applicable to the national forests, in such a manner as will best provide for—"
    - "(1) public recreation and enjoyment;"
    - "(2) the conservation and development of the scenic, natural, historic, and pastoral values of the area;"
    - "(3) the management, utilization, and disposal of natural resources such as timber, grazing, and mineral resources so that their utilization will not substantially impair the purposes; and"
    - "(4) the management of water quality ... consistent with the development of needed water supply and waste-water systems,

including the control of aquatic vegetation in the streams, lakes, and reservoirs."

- §4(c)—requires "an overall management plan," in consultation with state and local governments and interested persons.
- §5—authorizes land and easement acquisition.
- §6—permits hunting and fishing. Allows zones and periods of no hunting or fishing for public safety, area general administration, or public use and enjoyment; requires consultation with state.
- §7—requires cooperation on issuing permits and development and operation of facilities and services.
- §8—specifies no "express or implied claim or denial on the part of the Federal Government as to exemption from State water laws."
- §10—has no effect on state or local rights "to exercise civil and criminal jurisdictions within the recreation area or of its right to tax persons, corporations, franchises, or property."
- §11—authorizes appropriations.
- §12—designates wilderness.

9. Rattlesnake National Recreation Area (MT)—P.L. 96-476, Rattlesnake National Recreation Area and Wilderness Act of 1980; 21/2 pages.

- §1(a)—presents findings that "certain other lands ... while not predominantly of wilderness quality, have high value for municipal watershed, recreation, wildlife habitat, and ecological and educational purposes."
- §1(b)—establishes purposes: "to promote the watershed, recreational, wildlife, and educational values."
- §2—designates wilderness.
- §3—designates recreation area.
- §4—authorizes (a) land acquisition; and (b) land exchange.
- §4(c)—directs treatment of coal leasing rights.
- §4(d)(2)—directs that "nothing ... shall ... affect or diminish any water right ... nor the rights of the owner of such water right to the customary and usual access, including necessary motorized use ... to operate and maintain such facilities."
- §6—authorizes appropriations.

10. Robert T. Stafford White Rocks National Recreation Area (VT)—P.L. 98-322, Vermont Wilderness Act of 1984, Title II; 2 pages.

- §201(a)—presents findings.
- §201(b)—establishes purposes: "to preserve and protect their existing wilderness and wild values and to promote wild forest and aquatic habitat for wildlife, watershed protection, opportunities for primitive and semiprimitive recreation, and scenic, ecological, and scientific values."
- §202—designates 36,400-acre area.
- §204(a)—directs management, subject to valid existing rights, "in accordance with ... the laws, rules, and regulations applicable to the national forests" for:
    - "(1) the continuation of existing primitive and semiprimitive recreational use in a natural environment;"
    - "(2) utilization of natural resources ... only if consistent with the findings and purposes of this title;"
    - "(3) preservation and protection of forest and aquatic habitat for fish and wildlife; and"
    - "(4) protection and conservation of special areas having uncommon or outstanding wilderness, biological, geological, recreational, cultural, historical or archeological, and scientific, or other values contributing to public benefit."
- (b)—withdraws area from mineral leasing laws.
- (c)—permits hunting, fishing, and trapping.
- (d) & (e)—requires within 18 months a comprehensive management plan, and "shall provide for full public participation, shall consider the views of all interested agencies, organizations, and individuals, and shall particularly emphasize the values enumerated" above.

11. Oregon Cascades Recreation Area (OR)—P.L. 98-328, Oregon Wilderness Act of 1984, §4; 2+ pages.

- (a)—purposes are "to conserve, protect, and manage, in a substantially undeveloped condition, certain ... lands ... having unique geographic, topographic, biological, ecological features and possessing significant scenic, wildlife, dispersed recreation, and watershed values."
- (b)—designates 156,900-acre area; directs management "in accordance with the laws and regulations applicable to the National Forest System so as to enhance scenic and watershed values, wildlife habitat, and dispersed recreation."

- (c)—requires plans to:
  - "(1) provide a range of recreation opportunities from primitive to full service developed campgrounds;"
  - "(2) provide access for use by the public;"
  - "(3) to the extent practicable, maintain the natural and scenic character of the area; and"
  - "(4) provide for the use of motorized recreation vehicles."
- (d)—(1) directs that mining claims are subject to "reasonable regulations"; (2) withdraws area from mining and mineral leasing laws, subject to valid existing rights.
- (e)—permits "under appropriate regulations ... for resource protection and management and for visitor safety and comfort."
  - "(1) those [activities] necessary to prevent and control wildfire, insects, diseases, soil erosion, and other damaging agents, including timber harvesting activities necessary to prevent catastrophic mortality from insects, diseases or fire;"
  - "(2) those necessary to maintain or improve wildlife habitat, water yield and quality, forage production, and dispersed outdoor recreation opportunities;"
  - "(3) livestock grazing, to the extent that such use will not significantly adversely affect the resources of the recreation area;"
  - "(4) salvage of major timber mortality caused by fire, insects, disease, blowdown, or other causes when the scenic characteristics of the recreation area are significantly affected, or the health and safety of the public is threatened, or the overall protection of the forested area inside or outside the recreation area might be adversely affected by failure to remove the dead or damaged timber;"
  - "(5) those developments or facilities necessary for the public enjoyment and use of the recreation area, and"
  - "(6) public service land occupancies, including power transmission lines, provided there is no feasible alternative location."
- (f)—designates wilderness.
- (g)—requires either NFMA plan amendments for the four affected national forests or an integrated plan for the area within three years. "Any plan ... for the recreation area shall identify and designate

specific and appropriate areas and routes for the use of motorized recreational vehicles."

12. Mount Baker National Recreation Area (WA)—P.L. 98-339, Washington State Wilderness Act of 1984 §7; 1/2 page.

- (a)—establishes purposes: "to assure the conservation and protection of certain natural, scenic, historic, pastoral, and fish and wildlife values and to provide for the enhancement of the recreational values associated therewith."
- (b)—designates 8,600-acre area.
- (d)—directs management "in accordance with the laws, rules and regulations applicable to the national forests ... as will best provide for"
  - "(1) public outdoor recreation (including but not limited to snowmobile use);"
  - "(2) conservation of scenic, natural, historic, and other values contributing to public enjoyment; and"
  - "(3) such management, utilization, and disposal of natural resources ... which are compatible with and which do not significantly impair the purposes."

13. Allegheny National Recreation Area (PA)—P.L. 98-585, Pennsylvania Wilderness Act of 1984, §6; 1+ pages.

- (a)—designates 23,100-acre area.
- (b)—directs management "for the purposes of—"
  - "(1) outdoor recreation including, but not limited to, hunting, fishing, hiking, backpacking, camping, nature study, and the use of motorized and nonmotorized boats on the Allegheny Reservoir;"
  - "(2) the conservation of fish and wildlife populations and habitats;"
  - "(3) the protection of watersheds and the maintenance of free flowing streams and the quality of ground and surface waters;"
  - "(4) the conservation of scenic, cultural, and other natural values of the area;"
  - "(5) allowing the development of privately owned gas, oil, and mineral resources subject to reasonable regulations; and"

- "(6) minimizing, to the extent practicable, environmental disturbances caused by resource development, consistent with the exercise of private property rights."
- (c)—directs management "in accordance with the laws, rules, and regulations applicable to the National Forest System." Provides that "subject to valid existing rights, any activity associated with the exploration, development, or transportation of oil, gas, or other minerals shall be subject to such reasonable regulations.... For any such activity, the Secretary shall require a plan of operations which shall including provisions for adequate reclamation, including, to the extent practicable, revegetation and rehabilitation after each phase of operations."
- (d)—requires a management plan for the area, "in conjunction with, or incorporated with" the NFMA plan for the Allegheny National Forest.
- (e)—permits hunting, fishing, and trapping. Allows zones and periods of no hunting, fishing, or trapping for public safety, administration, or public use and enjoyment; requires consultation with state.
- (f)—withdraws area from mining and mineral leasing laws, subject to valid existing rights.

14. Pine Ridge National Recreation Area (NE)—P.L. 99-504, Nebraska Wilderness Act of 1986, Title II; 11/2 pages.

- §201—designates 6,600-acre area.
- §203(a)—directs management, subject to valid existing rights, "in accordance with the laws, rules, and regulations applicable to the national forests ... [for]:"
  - "(1) the continuation of existing primitive and semiprimitive recreational use in a natural environment;"
  - "(2) preservation and protection of forest, aquatic and grassland habitat;"
  - "(3) protection and conservation of special areas having uncommon or outstanding wilderness, biological, geological, recreational, cultural, historical or archeological, and scientific, or other values contributing to the public benefit;"
  - "(4) the continuation of existing livestock grazing uses;"
  - "(5) the control of noxious weeds and insects and prevention of their spreading onto the nearby private and Federal lands; and"

  - "(6) the control of fires and prevention of their spreading onto nearby private and Federal lands."
- §203(b)—requires a Memorandum of Understanding (MOU) "with local and State firefighting agencies and individuals to assure the best utilization of the firefighting resources available."
- §203(c)—permits hunting, fishing, and trapping.
- §203(d)—withdraws area from mining and mineral leasing laws, subject to valid existing rights.
- §203(e)—has no effect on state fish and wildlife jurisdictions or responsibilities.
- §203(f) & (g)—requires a comprehensive management plan within 18 months, with "full public participation."

15. Winding Stair Mountain National Recreation Area (OK)—P.L. 100-499, Winding Stair Mountain National Recreation Area and Wilderness Act, §8; 1 page

- (a)—establishes purposes: "to ensure the conservation and protection of certain natural, scenic, historic, pastoral, and fish and wildlife values and to provide for the enhancement of the recreational values associated therewith."
- (b)—designates 26,445-acre area.
- (d)—directs management "in accordance with the laws, rules and regulations applicable to the national forests.... Management and utilization of natural resources ... shall be permitted to the extent such management and utilization is compatible with and does not impair the purposes."
- (e)—directs that "any sales of timber ... shall be designed so as to not detract from the scenic values.... Management practices that would detract from the scenic quality and natural beauty within the view from [two roads] ... shall not be conducted.... Unevenaged timber management shall be the timber management practice in the recreation area, except that the Secretary may use evenaged management practices in order to promote public safety, mitigate the effects of fire, insects, and disease, or allow scenic vistas and recreational development or if such practices result in irregular cuts behind geographic barriers blocking the view from" two roads.[33]

16. Grand Island National Recreation Area (MI)—P.L. 101-292; 6+ pages.

- §1—designates area "to preserve and protect ... the outstanding resources and values ... [and for] providing the for conservation, protection, and enhancement of its scenery, recreation, fish and wildlife, vegetation and historical and cultural resources."
- §2—defines area boundaries.
- §3(a)—directs management "in accordance with the laws, rules, and regulations applicable to the National Forest System."
- §3(b)—establishes additional management guidance:
  - (1)—provides access across federal lands to private homes and lands, "subject to ... terms and conditions."
  - (2)—requires provision and maintenance of "traditional public access, including vehicular roads for general recreational activities, such as camping, hiking, hunting, fishing, and trapping."
  - (3)—permits "the use of snowmobiles ... regulated to protect the resources [to] ... minimize the degradation of these resources."
  - (4)—allows "timber management ... only as a tool to enhance public recreation, scenic quality, game and nongame wildlife species, and the protection and enhancement of threatened, endangered, or sensitive species. Trees damaged or downed due to fire, insects, disease, or blowdown may be utilized, salvaged, or removed."
  - (5)—provides for "reasonable water transportation from the mainland ... [via] concession, permit, or other means, and a reasonable charge may be imposed."
  - (6)—directs "through concession, permit, or other means docking and lodge facilities."
  - (7)—directs "reasonable actions to provide for public health and safety and for the protection of the ... area in the event of fire or infestation of insects or disease."
  - (8)—requires "as a condition of acquisition ... occupancy and use permits for any privately owned home ... within the ... area" for 20 years, and renewable, if "in compliance with the purposes of this Act, the terms of the permit, and other applicable rules and regulations," with specific provisions:

- (A)—permits are limited to the owner, "spouse, the children, stepchildren, and grandchildren ... and their direct lineal descendents (natural and adopted offspring)."
- (B)—use is restricted to "noncommercial recreation occupancy."
- (C)—annual fees are based on fair market value.
- (D)—"expansion, remodeling, or reconstruction ... [is] subject to approval" and limited in extent.
- (E)—federal acquisition is authorized at the fair market value.
- (F)—permits may be terminated for failure to comply with terms and conditions.
- (G)—improvements not removed within one year of termination belong to the federal government.

- §4—authorizes acquisition of lands, waters, structures, and interests.
- §5—has no effect on state fish and wildlife responsibility. Allows zones and periods of no hunting, fishing, or trapping for public safety, administration, protection of nongame species and their habitats, or public use and enjoyment; requires consultation with state.
- §6—withdraws area from mining and mineral leasing laws, except for "common varieties of mineral materials" subject to valid existing rights.
- §7—requires, when more than 10,000 acres have been acquired, a comprehensive management plan, within 30 months. Special issues to consider include:
    - "(1) Public recreation, including consideration of a range of appropriate recreational opportunities consistent with the rustic, natural, and historic character of the island."
    - "(2) The feasibility [and economics] of a concessionaire constructed, operated, and maintained rustic lodge and educational facility on no more than 55 acres" and with no effect on scenic views or tree line from Pictured Rocks National Lakeshore
    - "(3) Prescriptions concerning any management and harvest of timber."
    - "(4) General design criteria for new facilities or the improvement of existing facilities that are compatible with the rustic, natural, and historic character of the island and their topographic and

geological location, and that do not impair scenic views from Pictured Rocks."

- "(5) Water transportation from the mainland ... by a concessionaire or other entity."
- "(6) The feasibility of concessionaire constructed, operated, and maintained docking facilities in the natl. recreation area and on the mainland."
- "(7) An inventory and assessment of existing traditional roads, the level of road use, access needs, and any vehicular regulation and management needed to protect the resources ... while providing reasonable access to private property."

- §8—creates an advisory commission for advice on management plan; defines membership; and terminates the commission after the plan is adopted.
- §9—authorizes appropriations.

17. Smith River National Recreation Area (CA)—P.L. 101-612, 10 pages (excluding 4+ pages designating wild and scenic rivers).

- §2—presents findings.
- §3—includes definitions.
- §4—(a) establishes purposes: for "ensuring the preservation, protection, enhancement, and interpretation ... of the Smith River watershed's outstanding wild and scenic rivers, ecological diversity, and recreational opportunities while providing for the wise use and sustained productivity of its natural resources." (b) establishes boundaries.
- §5(a)—directs management "in accordance with ... the laws, rules, and regulations applicable to the National Forest System." The USFS "shall ... undertake [to]:"
    - "(1) Provide for a broad range of recreation uses and provide recreational and interpretive services and facilities."
    - "(2) Provide and maintain adequate public access."
    - "(3) Improve the anadromous fishery and water quality, including (but not limited to) stabilizing landslides, improving fish spawning and rearing habitat, and placing appropriate restrictions or limitations on soil disturbing activities."
    - "(4) Permit the use of off-road vehicles only on designated routes."

- "(5) Provide for public health and safety and for the protection of the recreation area in the event of fire or infestation of insects or disease."
- "(6) Permit programmed timber harvest only in those management areas where timber harvest is specifically authorized.... Timber management ... shall incorporate the use of strategies to reduce habitat fragmentation and employ silvicultural prescriptions designed to maintain or enhance biological diversity and wildlife habitats (such as retention of standing green trees, snags, and other coarse woody debris) by providing a high level of structural and compositional diversity in managed stands."
- "(7) Permit removal of trees within streamside protection zones ... [of rivers listed in §11] only when necessary for human health and safety, to maintain trails or existing roads, for the development of recreation or other facilities, for the protection of the recreation area in the event of fire, or to improve fish and wildlife habitat."
- "(8) permit removal of trees in ... areas where timber harvest is not specifically authorized ... when necessary for human health and safety, to maintain trails or existing roads, for the development of recreation or other facilities, for the protection of the recreation area in the event of fire, or to improve fish and wildlife habitat. Timber damaged or down in these areas as a result of fire, insects, disease, blowdown or other natural events shall otherwise be retained in its natural condition, with removal permitted only upon a written determination ... that such removal is necessary to provide for or maintain or enhance biological and ecological diversity, without regard to the commodity value of the timber."
- "(9) Provide for long-term viability and presence of Port-Orford-cedar [a particular tree species] and ensure its continued present economic and noneconomic uses."
- "(10) Except where timber harvest is specifically authorized ... protect, preserve, and increase old growth forest habitat."
- "(11) Provide for the restoration of landscapes damaged by past human activity."
- "(12) Develop a monitoring program to consistently gather water quality, air quality, wildlife, and fisheries data."

- "(13) Develop and implement a management plan to maintain, protect, and promote habitat for native resident trout species."
- "(14) Cooperate with other Federal, State, and local government agencies."

- §5(b)—notes eight management units within the recreation area, as defined on the referenced map, managed consistent with §5(a) plus additional management emphasis:
  - (A) for #1 "shall be on back-country and whitewater recreation, while recognizing unique botanic communities, outstanding whitewater, and historic and scenic values."
  - (B) for #2 "shall be on providing and maintaining ecologic and biologic diversity. Timber harvest shall be permitted ... only in existing plantations."
  - (C) for #3 "shall be on maintaining wildlife values and providing for a full range of recreation uses, with particular emphasis on scenic and recreation values associated. with the Smith Rivers, old growth redwoods, and Cal. State Hwy. 199."
  - (D) for #4 "shall be on wild rivers and roadless back-country recreation."
  - (E) for #5 "shall be on maintaining and protecting natural scenic values in the rivers canyon while providing for traditional and compatible rivers sports, including white water rafting, angling, sightseeing, and developed and dispersed recreation Timber harvests based on uneven-aged management with extended rotations shall be allowed where consistent with protection of the scenic values."
  - (F) for #6 "shall be on maintenance of wildlife values while providing rustic family and group recreation facilities for fishing, swimming, hunting, and camping. Timber harvests based on uneven-aged management with extended rotations shall be allowed where consistent with protection of scenic and wildlife values."
  - (G) for [#7] "shall be on providing a sustained yield of wood products while maintaining biological and ecological diversity."
  - (H) #8 is the Siskiyou Wilderness, and managed accordingly.
- §6—authorizes land acquisition and certain land transfers [1 page]; (d) withdraws area from disposal under public land laws, subject to valid existing rights.

- §7—has no effect on state jurisdiction or responsibilities over fish and wildlife. Allows zones and periods of no hunting, fishing, or trapping for protecting nongame species and their habitats, public safety, administration, or public use and enjoyment; requires consultation with state.
- §8—(a) withdraws area from mining and mineral leasing laws, subject to valid existing rights; (b) prohibits mining patents; (c) prohibits mineral development, except common variety minerals and subject to valid existing rights.
- §9—requires revision of existing management plan for area, to be "incorporated in its entirety" in the NFMA plan for the Six Rivers National Forest.
- §10—designates wild and scenic rivers—4+ pages.
- §11(a)—establishes streamside protection zones for listed streams [nearly 1 page]; "timber harvesting shall be prohibited" except under §5(a)(7). "Such zone shall extend 300 feet from each bank of the rivers and rivers segments, or 100 feet from the inner gorge, or within the limit of high and extreme landslide hazards, whichever is greater."
- §11(b)—establishes streamside protection zones for listed streams [nearly 1 page], "timber harvesting shall be prohibited" except under §5(a)(7). "Such zone shall extend on the average of one-quarter mile on either side of said rivers and rivers segments, or 100 feet from the inner gorge, or within the limit of high and extreme landslide hazards, whichever is greater."
- §12(a)—has no effect on state civil or criminal jurisdiction, state responsibility over fish and wildlife, or state authority "to tax persons, franchise, or public property."
- §12(b)—directs cooperative agreements for rescue, fire, law enforcement; and for nonfederal land planning within the area.
- §12(c)—authorizes technical assistance to state and local governments.
- §12(d)—directs assistance to a county in developing a land information system, compatible with federal systems.
- §13—has no effect on (a) lands outside the area; (b) prior rights, including timber contracts; (c) state road easements; (d) access by private landowners.
- §13(e)—directs 11 years of additional payments to states for lands in the county.
- §14—authorizes appropriations.

18. Ed Jenkins National Recreation Area (GA)—P.L. 102-217, Chattahoochee National Forest Protection Act of 1991, §4; 1 page. Originally named Springer Mountain National Recreation Area; name changed in P.L. 102-456.

- (a)—designates 23,330-acre area for "the protection of certain natural, scenic, fish and wildlife, historic and archaeological, wildland and watershed values, and providing for the enhancement of the recreation opportunities associated. with these values."
- (b)(1)—directs management, subject to valid existing rights, "in accordance with the laws, rules and regulations applicable to the national forests." Prohibits timber harvesting, except "in furtherance of this section, but only in a manner which does not impart the purposes.... Salvage or sanitation harvesting of timber stands which are substantially damaged by fire, windthrow or other catastrophe, or are in imminent danger from insect or disease attack, is authorized to maintain forest health. Timber harvesting is authorized to provide for visitor safety."
- (b)(2)—allows "the completion of existing timber sales under contract ... [and] additional road construction ... in furtherance of the purposes."
- (b)(3)—specifies no need to "change patterns of public access or closure on existing permanent national forest development roads ... [and] may open or close such existing roads to public use for reasons of sound resource management."
- (b)(4)—withdraws area from "all laws pertaining to mineral leasing."
- (b)(5)—allows "the continued maintenance of existing wildlife openings, ... [and] new wildlife openings in furtherance of the purposes."
- (b)(6)—requires management to "protect, enhance, and promote the public's opportunities for primitive and semiprimitive recreation."
- (b)(7)—prohibits interference "with the rights of access to privately held lands."

19. Spring Mountains National Recreation Area (NV)—P.L. 103-63; 4+ pages.

- §2—provides definitions.
- §3—establishes purposes, to:

  - "(1) preserve scenic, scientific, historic, cultural, natural, wilderness, watershed, riparian, wildlife, threatened and endangered species, and other values contributing to public enjoyment and biological diversity;"
  - "(2) ensure appropriate conservation and management of natural and recreation resources; and"
  - "(3) provide for the development of public recreation opportunities ... for the enjoyment of present and future generations."
- §4—designates 316,000-acre area.
- §5(a)—directs management "in accordance with the laws, rules, and regulations applicable to the National Forest System and ... for—"
  - "(1) the conservation of scenic, scientific, historic, cultural, and other values contributing to public enjoyment;"
  - "(2) the conservation of fish and wildlife populations and habitat, including the use of prescribed fire to improve or maintain habitat;"
  - "(3) the protection of watersheds and the maintenance of free flowing streams and the quality of ground and surface waters;"
  - "(4) public outdoor recreation benefits, including, but not limited to, hunting, fishing, trapping, hiking, horseback riding, backpacking, rock climbing, camping, and nature study;"
  - "(6) the management and use of natural resources in a manner compatible with the purposes."
- §5(b)—permits hunting, trapping, fishing, and habitat management. Allows zones and periods of no hunting, trapping, or fishing for public safety, administration, or public use and enjoyment; requires consultation with state.
- §5(c)—authorizes continued "grazing of livestock ... subject to such reasonable regulations, policies, and practices as ... necessary."
- §5(d)—authorizes "such reasonable measures as ... necessary to protect the land and resources from fire or insect or disease infestation."
- §6(a)—requires a management plan as an amendment to the NFMA plan for the Toiyabe National Forest within 3 full fiscal years and in consultation with state agencies; (2) identifies contents:
  - "(A) implementation plans for a continuing program of interpretation and public education;"

- "(B) proposals for public facilities to be developed, expanded, or improved;"
- "(C) plans for the management of natural and cultural resources, with emphasis on the preservation and long-term scientific use of archaeological resources;"
- "(D) wildlife and fish resource management plans ... in consultation" with the state;
- "(E) recreation management plans ... in consultation" with the state;
- "(F) wild horse and burro herd management plans ... in consultation" with the state; and
- "(G) an inventory of all lands ... that will permit ... possible future acquisitions."

- §6(b)—requires plan to include wilderness study area recommendations; requires management to maintain wilderness suitability "until otherwise directed by Act of Congress."
- §7—authorizes land acquisition.
- §8—withdraws area from public land, mining, mineral leasing laws, except for a specified site, subject to valid existing rights.
- §9—directs, for "unified and cost-effective management and interpretation," cooperative agreements with other federal, state, local governments, and nonprofits for "management and interpretation of natural and cultural resources."
- §10—authorizes appropriations.

20. Fossil Ridge Recreation Area (CO)—P.L. 103-77, Colorado Wilderness Act of 1993, §5; 1 page.

- (a)—designates 43,900-acre area "to conserve, protect, and enhance scenic, wildlife, recreational, and other natural resource values."
- (b)—directs management "in accordance with the laws and regulations applicable to the National Forest System."
- (c)—withdraws area from public land, mining, and mineral leasing laws, subject to valid existing rights.
- (d)—prohibits "timber harvesting ... except to the extent that would be permitted in wilderness ... for necessary control of fire, insects, and diseases, and for public safety."
- (e)—directs that "the designation ... shall not be construed to prohibit, or change the administration of, the grazing of livestock."

- (f)—directs that "no developed campgrounds ... [and] no new roads or trails may be constructed."
- (g)—allows "motorized travel ... only on those established trails and routes existing as of July 1, 1991.... Nothing ... shall be construed as precluding ... closing of any trail or route from use for ... resource protection or public safety."

21. Jemez National Recreational Area (NM)—P.L. 103-104; 41/2 pages.

- §1—designates 57,000-acre area "to conserve, protect, and restore the recreational, ecological, cultural, religious, and wildlife resource values."
- §2(a)—directs management "in accordance with ... the laws, rules, and regulations applicable to National Forest System lands.... Management of natural resources ... shall be permitted only to the extent ... compatible with and does not impair the purposes.... Recreational activities shall include (but not be limited to) hiking, camping, hunting, fishing, skiing, backpacking, rock climbing, and swimming."
- §2(b)—requires a management plan as an amendment to the NFMA plan for the Santa Fe National Forest within five years.
- §2(c)—requires "particular emphasis to the preservation, stabilization, and protection of cultural resources."
- §2(d)—"In recognition of the historic use ... by Indian peoples for traditional cultural and customary uses," requires consultation with local tribal leaders.
- §2(e)—requires "particular emphasis to the conservation and protection of wildlife resources, including species listed as sensitive."
- §2(f)—permits hunting and fishing.
- §2(g)—permits "timber harvesting ... for commercial purposes, including (but not limited to) vigas, latillas, the gathering of fuelwood, and for purposes of public safety, recreation, wildlife, and Administration, insofar as the harvesting is compatible with the purposes .... Trees damaged or downed due to fire, disease, or insect infestation may be utilized, salvaged, or removed ... in furtherance of the purposes.... Nothing ... shall be construed to affect the timber sales under contract." One particular timber sale is required to use "uneven aged management including the individual tree selection method."

- §2(h)—permits livestock grazing; "Riparian areas shall be managed ... to protect their important resource values."
- §2(i)—(1) requires, within one year, a transportation plan "for most efficient use of roads and trails ... [and] for a comprehensive trails system that provides for dispersed recreation while minimizing impact on significant archaeological and religions sites." (2) authorizes construction, maintenance, and closure of roads only after consultation with local tribal leaders.
- §2(j)—requires "recreation facilities ... constructed so as to minimize impacts on the scenic beauty, the natural character, and the archaeological and religious sites."
- §2(k)—requires "a visitor center and interpretive facilities."
- §2(l)—allows "a utility corridor for high power electric transmission lines ... only when ... (1) there is not a feasible alternative; (2) damage to the recreational and scenic quality and to the archaeological and religious sites will not be significant; (3) it is in the public interest; and (4) a plan to minimize harm to the resources ... has been developed."
- §2(m)—allows "scientific investigations ... in the public interest and ... compatible with the purposes."
- §2(n)—allows "zones where, and ... periods when, any activity otherwise permitted ... will not be permitted for reasons of public safety, administration, fish and wildlife management, protection of archaeological or cultural resources, or public use and enjoyment." Requires consultation with states and tribes, except in emergencies.
- §3—(a) prohibits mining patents; requires any claims of injury within one year; (b) withdraws area from mining, mineral leasing, and mineral materials disposal laws, subject to valid existing rights; (c) prohibits "surface disturbance ... except in accordance with requirements, including ... reasonable reclamation of disturbed lands;" (d) requires a review of mining claim validity within three years; (e) authorizes use of "mineral materials ... for public purposes."
- §4—allows evaluation of "lands adjoining ... for possible inclusion ... [with] recommendations to Congress."
- §5—authorizes land acquisition.
- §6—authorizes appropriations.

22. Opal Creek Scenic Recreation Area (OR)—P.L. 104-208, Omnibus Consolidated Appropriations Act, 1997; Division B—Oregon Resource

Conservation Act of 1996; Title I— Opal Creek Wilderness and Scenic Recreation Area; 6 pages. [Also enacted in P.L. 104-333, Omnibus Parks and Public Lands Management Act of 1996, §1023.]

- §102—provides definitions.
- §103—establishes purposes:
  - "(1) to protect and provide for the enhancement of the natural, scenic, recreational, historic, and cultural resources;"
  - "(2) to protect and support the economy of the communities ... ; and"
  - "(3) to provide increased protection for an important drinking water source."
- §104—designates 13,000-acre area, conditional upon specified land and easement donations and binding agreement on other interests.
- §105(a)—directs management "in accordance with this title and the laws (including regulations) applicable to the National Forest System."
- §105(b)—requires a management plan within two years to become part of the NFMA plan for the Willamette National Forest "and supersede any conflicting provision in such" plan.
- §105(c)—recognizes recreation as an appropriate use, permitted at current use levels, and increased if consistent with protecting resource values.
- §105(d)—requires a transportation plan; specifies guidelines.
- §105(e)—permits hunting and fishing. Allows zones and periods of no hunting of fishing for public safety, administration, or public use and enjoyment; requires consultation with state.
- §105(f)—(1) generally prohibits "the cutting and/or selling of trees." (2)(A) allows "the cutting of trees ... only—(i) for public safety, such as to control the continued spread of a forest fire; (ii) for activities related to administration; or (iii) for removal of hazard trees along trails and roadways." (2)(B) prohibits salvage sales.
- §105(g)—withdraws area from public land, mining, and mineral leasing laws, subject to valid existing rights.
- §105(h)—provides details on administration of Bornite project (a mineral project within the area).
- §105(i)—specifies that FERC "may not license the construction of any dam, water conduit, reservoir, powerhouse, transmission line, or other project work" except for Bornite project.

- §105(j)—requires a review and revision of the cultural and historic resource inventory within one year.
- §105(k)—requires (1) consultation with the advisory council; (2) views of private groups, individuals, and the public; (3) views and assistance from other government agencies; and (4) views of nonprofits.
- §106—creates an advisory council; specifies membership.
- §107—(a) authorizes land acquisition; (b) has no effect on environmental response and cost recovery (i.e., Superfund) actions; (d) does not interfere with special use permits.

23. Land Between the Lakes National Recreation Area (KY-TN)—P.L. 105-277, Omnibus Consolidated and Emergency Supplemental Appropriations, 1999, Div. A, §101(e), Title V; 2 pages (6 columns) in the Congressional Record.

- §502—provides definitions.
- §503—establishes purposes: "(1) to transfer jurisdiction without consideration administrative jurisdiction; (2) to protect and manage the resources ... for optimum yield of outdoor recreation and environmental education through multiple use management; (3) to authorize, research, test, and demonstrate innovative programs and cost-effective management; (4) to authorize ... to cooperate ... and to help stimulate the development of the surrounding region and extend the beneficial results as widely as practicable; and (5) to provide for the smooth and equitable transfer."
- §511(b)—directs management (1) "for multiple use as a unit of the" NFS; (2) "The emphasis ... shall be—(A) to provide public recreational opportunities; (B) to conserve fish and wildlife and their habitat; and (C) to provide for diversity of native and desirable non-native plants, animals, opportunities for hunting and fishing, and environmental education."
- §512—extends Tennessee Valley Authority (TVA) law enforcement authorities to Forest Service (FS) agents and officers.
- §513—includes area for Payments In Lieu of Taxes and USFS payments to states; continues TVA payments.
- §514—specifies, for transportation funding, that all paved roads are forest highways.
- §521—requires a management plan under NFMA.

- §522—creates an advisory board; specifies membership, terms, functions, and permanence.
- §523—(a) authorizes "reasonable fees for admission to and the use of the designated sites, or for activities;" (b) requires consideration of "the costs of collection weighed against potential income;" (c) prohibits "general entrance fees."
- §524—requires fees deposited into a special fund, available without further appropriation for management.
- §525—provides additional authority for special use authorizations.
- §526(a)—allows a special use authorization for U.S. Fish and Wildlife Service facilities, with fees deposited into the special fund, cooperation between the agencies, and fish and wildlife activities subordinate to overall management.
- §526(b)—allows grants, contracts, cooperative agreements, and gifts for "the management, maintenance, operation, and interpretation."
- §527—designates a national recreation trail.
- §528—requires "an inventory of and ... access to cemeteries ... for ... burial, visitation, and maintenance."
- §529(a)—withdraws area from mining and mineral leasing laws; "may permit the use of common varieties of mineral materials."
- §529(b)—(1) permits hunting and fishing; (2) allows zones and periods of no hunting or fishing for public safety, administration, or public use and enjoyment; requires consultation with state; (3) has no effect on state wildlife and fish jurisdiction or responsibilities.
- §530—requires TVA maintenance/repair of Hematite Dam for a year, then transfers responsibility to USDA.
- §531—establishes a trust fund, with $1 million annually for five years from TVA, for "(1) public education, grants, and internships related to recreation, conservation, and multiple use land management ... and (2) regional promotion ... in cooperation with development districts, chambers of commerce, and State and local governments."
- Subtitle C—Transfer Provisions includes 2 pages (6 columns) in the *Congressional Record* on transfers of personnel, property, records, responsibilities, and funding.
- Subtitle D—Funding; §551 authorizes appropriations.

24. Moosalamoo National Recreation Area (VT)—P.L. 109-382, New England Wilderness Act of 2006, Title II—Vermont, Subtitle B; 1 page.

- §221—designates 15,857-acre area.
- §223(a)—directs management "in accordance with—(1) laws (including rules and regulations) applicable to units of the National Forest System; and (2) the management direction (including objectives, standards, and guidelines) established for the Moosalamoo Recreation and Education Management Area under the" existing NFMA plan.
- §223(b)—has no effect on state fish and wildlife jurisdiction.
- §223(c)—has no effect on "managing the Green Mountain Escarpment Management Area and the Ecological Special Areas," in the NFMA plan.

25. Mount Hood National Recreation Area (OR)—P.L. 111-11, Omnibus Public Land Management Act of 2009, Title I, Subtitle C—Mount Hood Wilderness, Oregon, §1204; 2 pages.

- (a)—establishes purposes: "to provide for the protection, preservation, and enhancement of recreational, ecological, scenic, cultural, watershed, and fish and wildlife values."
- (b)—designates 34,550-acre area.
- (d)—directs management "in accordance with the laws (including regulations) and rules applicable to the National Forest System."
- (e)—permits "the cutting, sale, or removal of timber"
  - "(1) to the extent necessary to improve the health of the forest in a manner that—
    - "(A) maximizes the retention of large trees—(i) as appropriate to the forest type; and (ii) to the extent that the trees promote stands that are fire-resilient and healthy;"
    - "(B) improves the habitats of threatened, endangered, or sensitive species; or"
    - "(C) maintains or restores the composition and structure of the ecosystem by reducing the risk of uncharacteristic wildfire;"
  - "(2) to accomplish an approved management activity ... if the cutting, sale, or removal of timber is incidental; or"
  - "(3) for de minimus personal or administrative use."
- (f)—prohibits "new or temporary roads ... except as necessary—"

- "(1) to protect the health and safety of individuals in cases of an imminent threat of flood, fire, or any other catastrophic event that, without intervention, would cause the loss of life or property;"
- "(2) to conduct environmental cleanup;"
- "(3) to allow for the exercise of reserved or outstanding rights provided for by a statute or treaty;"
- "(4) to prevent irreparable resource damage by an existing road; or"
- "(5) to rectify a hazardous road condition."

- (g)—withdraws area from public land, mining, and mineral leasing laws, subject to valid existing rights.
- (h)—transfers jurisdiction of about 130 acres from the BLM to the Forest Service.

26. Bridgeport Winter Recreation Area (CA)—P.L. 111-11, Omnibus Public Land Management Act of 2009, Title I, Subtitle K—Eastern Sierra and Northern San Gabriel Wilderness, California, §1806; 1+page.

- (a)—designates 7,254-acrea area.
- (c)—(1) directs management under existing NFMA plan until completion of area plan required in §1806(d); and (2) allows snowmobile use "(A) during periods of adequate snow coverage; and (B) subject to any terms and conditions determined to be necessary."
- (d)—requires, within one year and through a public process, "a winter use management plan that provides for—"
  - "(1) adequate signage;"
  - "(2) a public education program on allowable usage areas;"
  - "(3) measures to ensure adequate sanitation;"
  - "(4) a monitoring and enforcement strategy; and"
  - "(5) measures to ensure the protection of the [Pacific Crest National Scenic] Trail."
- (e)—directs priorities for enforcement as"
  - "(1) to prohibit degradation of natural resources;"
  - "(2) to prevent interference with nonmotorized recreation on the Trail; and"
  - "(3) to reduce user conflicts in the Recreation Area."
- (f)—requires "an appropriate snowmobile crossing point along the [Pacific Crest National Scenic] Trail" in accordance with applicable laws and subject to terms and conditions "to ensure that the crossing

would not—(A) interfere with the nature and purposes of the Trail; or (B) harm the surrounding landscape."

Scenic Areas

Congress has designated 11 scenic areas within the National Forest System, each with distinct management direction enacted in the law establishing the area. More than half have been named *national scenic areas*, but several are unique variations: *scenic-research area*; *scenic highway*; *national forest scenic area*; and *national scenic and wildlife area*. The rationale for these distinctions is not apparent. No conclusions can be drawn about the general management guidance for these areas, or the differences compared to national monuments and recreation areas. In general, scenic areas are smaller than national monuments and recreation areas; most are less than 10,000 acres, but they range up to 128,303 acres for the Mono Basin National Forest Scenic Area.

1. Cascade Head Scenic-Research Area (OR)—P.L. 93-535; 31/2 pages.

- §1—designates area "to provide present and future generations with the use and enjoyment of certain ocean headlands, rivers, streams, estuaries, and forested areas, to insure the protection and encourage the study of significant areas for research and scientific purposes, and to promote a more sensitive relationship between man and his adjacent environment."
- §2—directs management "in accordance with the laws, rules, and regulations applicable to national forests."
- §3(c)—establishes subareas, with supplemental purposes:
  - "(1) Estuary and Associated Wetlands Subarea: to protect and perpetuate the fish and wildlife, scenic and research-education values, while allowing dispersed recreation use, such as sport fishing, nonmotorized pleasure boating, waterfowl hunting, and other uses ... compatible with the protection and perpetuation of the unique natural values. After appropriate study, breaching of dikes may be permitted."
  - "(2) Lower Slope-Dispersed Residential Subarea: to maintain the scenic, soil and watershed, and fish and wildlife values, while

allowing dispersed residential occupancy, selective recreation use, and agricultural use."

- "(3) Upper Timber Slope and Headlands Subareas: to protect the scenic, soil and watershed, and fish and wildlife values while allowing selective recreation and extensive research-educational activities. Timber harvesting activity may occur ... only when ... such harvesting is to be conducted in connection with research activities or that the preservation of the timber resource is imminently threatened by fire, old age, infestation, or similar natural occurrences."
- "(4) Coastline and Sand Dune-Spit Subareas: to protect and maintain the scenic and wildlife values while allowing selective recreation and extensive research-educational activities."

- §5—(a) generally authorizes acquisition of lands, values, or interests; (b) except for the estuary subarea and associated wetlands subarea, acquisition requires landowner consent, unless there is imminent danger of a change in use; (c) nonfederal landowners must give 30 days notice of substantial change in use or maintenance.
- §7—withdraws area from mining and mineral leasing laws, subject to valid existing rights.
- §8—creates an advisory committee, pursuant to Federal Advisory Committee Act.
- §9—directs cooperation with state and local governments; has no effect on the state's "right to exercise civil and criminal jurisdiction ... or of its right to tax persons, corporations, franchises or other non-Federal property."

2. North Cascades Scenic Highway (WA)—P.L. 98-339, Washington State Wilderness Act of 1984, §8; 1/2 page.

- (a)—purposes are "scenic beauty and recreation qualities."
- (b)—designates 87,757-acres area. "Management activities, including resource use and development, within the area may be permitted ... if the existing scenic values ... are maintained."

3. Mono Basin National Forest Scenic Area (CA)—P.L. 98-425, California Wilderness Act of 1984, Title III; 6 pages.

- §301—designates area.

- §303—authorizes land acquisition; authorizes acquisition without consent if a significant change in private development is proposed.
- §304(a)(1)—directs management "as a separate unit ... in accordance with the laws, rules, and regulations applicable to the National Forest System."
- §304(b)(1)—directs management "consistent with the protection of the water rights, ... to protect its geologic, ecologic, and cultural resources ... [and] provide for recreational use,... [compatible] recreational and interpretive facilities ... [and] full use ... for scientific study and research."
- §304(b)(2)—prohibits "commercial timber harvesting ... but ... shall permit the utilization of wood material such as firewood, posts, poles, and Christmas trees by individuals for their domestic purposes ... [and] may take action including the use of commercial timber harvest to the minimum extent necessary to control fires, insects and diseases that might (A) endanger irreplaceable features or (B) cause substantial damage to significant resources adjacent to the Scenic Area."
- §304(c)—allows valid grazing permits to continue.
- §304(d)—authorizes cooperative agreements with state and local governments to protect their resources.
- §304(e)—requires "a detailed and comprehensive management plan" within three years, protecting water rights and including "(1) an inventory of natural (including geologic) and cultural resources; (2) general development plans for public use facilities, including cost estimates; and (3) measures for the preservation of the natural and cultural resources." Allows hunting and fishing (including commercial brine shrimp operations) except as "necessary for ... public health and safety, the protection of resources, scientific research activities, or public use and enjoyment."
- §304(f)—authorizes construction of a visitor center.
- §304(g)(1)—withdraws area from mining, mineral leasing, public land laws, subject to valid existing rights.
- §304(g)(2)—provides, subject to valid existing rights, "all mining claims ... shall be subject to such reasonable regulations." Patents are limited to title to the minerals and surface occupancy.
- §304(h)—"Nothing in this title shall be construed to reserve any water ... or to affirm, deny, or otherwise affect the present (or prospective) water rights ... nor ... to cause, authorize, or allow any interference or infringement of such water rights."

- §304(i)—repeals 1936 law on sale of land, granting of rights-of-way. Authorizes grants and rights-of-way easements at no cost.
- §304(j)—allows "existing community recreational uses" to continue.
- §305—requires a National Academy of Sciences study of the area's ecology.
- §306—creates an advisory board; defines membership.
- §307—requires "nonexclusive access ... by Indian people for such traditional cultural and religious purposes."
- §308—authorizes appropriations.

4. Columbia River Gorge National Scenic Area (OR-WA)—P.L. 99-663, Columbia River Gorge National Scenic Area Act; 281/2 pages.

- §2—provides definitions.
- §3—establishes purposes: "(1) to ... protect and provide for the enhancement of the scenic, cultural, recreational, and natural resources; and (2) to protect and support the economy of the ... area by encouraging growth to occur in existing urban areas and by allowing future economic development in a manner that is consistent with paragraph (1)."
- §4—designates (a) national scenic area; (b) special management areas (SMAs) within area; (c) boundaries; (d) a special purchase unit; and (e) urban areas.
- §5—establishes a Commission to facilitate cooperation; defines membership.
- §6—requires a management plan from the Commission, via several steps:
    - (a)—complete, within one year, studies of (1) resource inventory, including "all existing land uses, natural features and limitations, scenic, natural, cultural, archaeological and recreation and economic resources and activities;" (2) economic opportunities; (3) recreation resources and opportunities;
    - (b)—complete, within two years, land use designations for nonfederal lands;
    - (c)—complete, within three years, a management plan.
    - (d)—the plan must include standards to:
        - "(1) protect and enhance agricultural lands ... and to allow ... conversion ... to open space, recreational development, or forest lands;"

    - "(2) protect and enhance forest lands ... and to allow ... conversion ... to agricultural lands, recreation development or open spaces;"
    - "(3) protect and enhance open spaces;"
    - "(4) protect and enhance public and private recreation resources and educational and interpretive facilities and opportunities;"
    - "(5) prohibit major development actions in special management areas;"
    - "(6) prohibit industrial development in the scenic area outside urban areas;"
    - "(7) require that commercial development outside urban areas take place without adversely affecting the scenic, cultural recreation, or natural resources;"
    - "(8) require that residential development outside urban areas take place without adversely affecting the scenic, cultural recreation, or natural resources; and"
    - "(9) require that the exploration, development and production of mineral resources, and the reclamation of the land thereafter, take place without adversely affecting the scenic, cultural recreation, or natural resources."
  - (e)—requires consultation with federal, state, local governments and with tribes; requires public hearings and comments.
  - (f)—requires Secretarial approval of Commission's management.
  - (g) & (h)—requires plan revision at least every 10 years, and amendments when conditions have significantly changed.
- §7—directs management of nonfederal lands within the scenic area by the Commission via county-specific land use ordinances.
- §8(a)—directs SMA management "in accordance with ... the laws, rules and regulations applicable to the national forest system." Requires "the construction of roads and the management, utilization and harvest of timber on Federal lands ... shall also be subject to Forest Service visual resource management guidelines."
- §8(b)—withdraws SMAs from public land, mining, and mineral leasing laws, subject to valid existing rights. Allows "the exploration, development, or production of sand, gravel, and crushed rock."
- §8(c)—requires a resource inventory.
- §8(d)—requires a recreation assessment, within two years, to:

    - "(1) identify areas ... for ... an interpretive center [in Oregon] and a conference center" in Washington;
    - "(2) identify areas ... suitable for other public use facilities, including but not limited to educational and interpretive facilities, campsites, picnic areas, boat launch facilities, and river access areas; and"
    - (3) given treaty rights, "identify areas ... suitable for use to increase access for recreation."
- §8(e)—requires, in consultation with the commission, land use ordinances for nonfederal lands within the SMAs within three years.
- §8(f)—specifies guidelines for land use ordinances. "The guidelines shall require that management, utilization, and disposal of timber, and exploration, development, and production of sand, gravel, and crushed rock for the construction, maintenance, or reconstruction of roads used to manage or harvest forest products on nonfederal lands within the SMAs take place without adversely affecting the scenic, cultural, recreation, and natural resources of the scenic area."
- §8(h)-(n)—specifies process and timing for adopting SMA land use ordinances, with review by the Commission and concurrence of the Secretary.
- §8(o)—directs that the ordinances shall not apply if a landowner's offer to sell at fair market value is still pending after three years.
- §9(a)—authorizes land acquisition.
- §9(b)—establishes limitations on use of eminent domain.
- §9(c)—provides for hardship cases.
- §9(d)—provides land exchange guidance and a list of available federal lands (3 pages).
- §10—provides interim management guidelines, pending land use ordinances.
- §11—requires Oregon and Washington to each have an economic development plan; provides $5 million for each state in grants and loans for economic development, with conditions; requires an annual report on use of the funds.
- §12—directs management of the Old Columbia River Highway, to "preserve and restore the continuity and historic integrity."
- §13(a) & (b)—directs water resource projects on Columbia River streams and tributaries be treated as on wild and scenic rivers, except along or through Indian reservations.

- §13(c) & (d)—designates two wild and scenic rivers; requires two wild and scenic river studies.
- §14(a)—provides nonreimbursable technical assistance to counties to develop land use ordinances.
- §14(b) & (c)—directs special USFS payments to states for lands in the scenic area.
- §14(e)—generally (with listed exceptions) prohibits federal expenditures and new or renewed licenses, permits, and exemptions if the Commission is not established within 15 months or is subsequently disestablished.
- §15—directs (a) administrative, and (b) judicial enforcement efforts.
- §16—authorizes appropriations.
- §17(a)—includes savings provisions, specifying no effect on (1) Indian treaties or rights; (2) "appropriation or use of water;" (3) water and groundwater rights or jurisdictions, or use of water for transportation; (4) "any water or water-related right;" (5) any existing interstate compact; (6) "the ability of the Bonneville Power Administration to operate, maintain, and modify existing transmission facilities;" (7) land held in trust for tribes or individuals; (8) "the laws, rules and regulations pertaining to hunting and fishing;" (9) any NFMA plan; (10) prohibits "protective perimeters or buffer zones.... The fact that activities or uses inconsistent ... can be seen or heard ... shall not, of itself, preclude such activities or uses up to the boundaries."
- §17(b)—has no effect on Corps of Engineers responsibilities to improve navigation facilities at Bonneville Dam, except for disposal of excavation material.
- §17(c)—has no effect on nonfederal timber rights and responsibilities, except in SMAs.
- §17(d)—directs that "Mandatory language ... respecting the powers and responsibilities of the Commission shall be interpreted as conditions precedent to congressional consent to the interstate compact."
- §17(e)—establishes no federal obligation if states fail to comply with provisions.
- §17(f)—defines numerous actions (identified by section) as not major federal actions, and thus not requiring an environmental impact statement or environmental assessment.
- §18—includes severability.

5. Indian Nations National Scenic and Wildlife Area (OK)—P.L. 100-499, Winding Stair Mountain National Recreation Area and Wilderness Act, §10; 1/2 page.

- (a)—designates area "to protect and enhance certain scenery and wildlife."
- (c)—directs management "in accordance with the laws, rules and regulations applicable to the national forests.... Management practices ... that would detract from the scenic quality and natural beauty of the ... viewsheds shall be prohibited. Timber management practices ... shall promote a mixed hardwood and conifer forest with species and age class diversity approximating natural succession and with significant mast production and den trees for wildlife. Unevenaged management shall be the timber management practice, except ... to promote public safety, mitigate the effects of fire, insects, and disease, or if such practices result in irregular cuts behind geographic barriers blocking the view from" certain roads.

6. Beech Creek National Scenic Area (OK)—P.L. 100-499, Winding Stair Mountain National Recreation Area and Wilderness Act, §11; 1/2 page.

- (a)—designates area "to protect and enhance certain scenery and wildlife."
- (c)—directs management "in accordance with the laws, rules, and regulations applicable to the national forests.... Timber management practices ... shall promote a mixed hardwood and conifer forest with species and age class diversity approximating natural succession and with significant mast production and den trees for wildlife. Unevenaged management shall be the timber management practice, except ... to promote public safety or to mitigate the effects of fire, insects, and disease."

*7. **Coosa Bald National Scenic Area*** (GA)—P.L. 102-217, Chattahoochee National Forest Protection Act of 1991, §3; 1 page.

- (a)—designates 7,100-acre area for "protecting and enhancing the natural beauty, special ecological features, watershed integrity, mature-forest habitat, scenic recreation opportunities, and other distinctive values."
- (b)(1)—directs management, subject to valid existing rights, "in accordance with the laws, rules, and regulations applicable to the National Forest System."
- (b)(2)—permits "additional road construction ... in furtherance of the purposes." Prohibits timber harvesting except "in furtherance of the purposes, but only in a manner which does not impair the purposes.... Salvage or sanitation harvesting of timber stands which are substantially damaged by fire, windthrow or other catastrophe, or are in imminent danger from insect or disease attack, is authorized to maintain forest health. Timber harvesting is authorized to provide for visitor safety."
- (b)(3)—authorizes retention of "patterns of public access or closure on existing ... roads ... [and] may open or close such existing roads for public use for reasons of sound resource management."
- (b)(4)—requires completion of existing timber sales under contract.
- (b)(5)—withdraws area from mineral leasing laws.
- (b)(6)—permits "the continued maintenance of existing wildlife openings ... and may permit new wildlife openings in furtherance of the purposes" in cooperation with the state and other cooperators.
- (b)(7)—directs management to "protect, enhance, and promote the public's opportunities for primitive and semiprimitive experiences."

8. Mount Pleasant National Scenic Area (VA)—P.L. 103-314, George Washington National Forest Mount Pleasant National Scenic Area Act; 2+ pages.

- §2—establishes purposes to:
  - "(1) ensure appropriate protection and preservation of the scenic quality, water quality, natural characteristics, and water resources;"
  - "(2) protect and manage vegetation to provide wildlife and fish habitat;"

  - "(3) provide areas that may develop characteristics of old-growth forests; and"
  - "(4) provide a variety of recreation opportunities."
- §3(a)—designates 7,580-acre area.
- §3(b)(1)—directs management "in accordance ... with the laws and regulations generally applicable to the National Forest System."
- §3(b)(2)—requires a management plan as an amendment to the NFMA plan for the George Washington National Forest within three years.
- §3(c)—prohibits "new permanent roads" except for access to private lands or interests.
- §3(d)—prohibits "timber harvest ... except as may be necessary in the control of fire, insects, and diseases and to provide for public safety and trail access.... May engage in vegetation manipulation practices for maintenance of existing wildlife clearings and visual quality. Firewood may be harvested for personal use along perimeter roads."
- §3(e)—allows motorized travel only on two authorized routes and "as necessary for administrative use ... and on temporary routes in support of wildlife management projects."
- §3(f)—directs that "wildfires shall be suppressed in a manner consistent with the purposes."
- §3(g)—allows control of "insect and disease outbreaks ... to maintain scenic quality, prevent tree mortality, reduce hazards to visitors, or protect private lands."
- §3(h)—directs management to "maintain or enhance existing water quality."
- §3(i)—withdraws area from mining and mineral leasing laws, subject to valid existing rights.

9. Saint Helena Island National Scenic Area (MI)—P.L. 106-431, Saint Helena Island National Scenic Area Act; 3 pages.

- §2(a)—establishes purposes:
  - "(1) to preserve and protect for present and future generations the outstanding resources and values; and"
  - "(2) to provide for the conservation, protection, and enhancement of primitive recreation opportunities, fish and wildlife habitat, vegetation, and historical and cultural resources of the island."
  - §2(b)—designates area.

- §3—establishes boundaries.
- §4(a)—directs management, subject to valid existing rights, "in accordance with the laws, rules, and regulations applicable to the National Forest System."
- §4(b)—requires, within three years of acquisition of 1/2 of authorized land, an amendment to the NFMA plan for the Hiawatha National Forest for area, including:
- (1) "Alternative means for providing public access from the mainland."
- (2) "no new permanent roads."
- (3) "No timber harvest ..., except as may be necessary in the control of fire, insects, and diseases, and to provide for public safety and trail access.... May engage in vegetation manipulation practices for maintenance of wildlife habitat and visual quality. Trees cut for these purposes may be utilized, salvaged, or removed."
- (4) "Motorized travel shall not be permitted ..., except on the waters of Lake Michigan, and as necessary for Administration use."
- (5) "Wildfires shall be suppressed in a manner consistent with the purposes of this Act, using such means as the Secretary deems appropriate."
- (6) "Insect and disease outbreaks may be controlled ... to maintain scenic quality, prevent tree mortality, or to reduce hazards to visitors."
- (7) "provide through concession, permit, or other means docking facilities."
- (8) "take reasonable actions to provide for public health and safety and for the protection of the scenic area in the event of fire or infestation of insects or disease."

- §4(c)—requires consultation with state and local governments; requires "full public participation, and consider the views of all."
- §5—has no effect on state jurisdiction or responsibilities for fish and wildlife.
- §6—withdraws area from mineral leasing laws and mineral development, except for common varieties of mineral materials, and subject to valid existing rights.
- §7—authorizes land acquisition.
- §8—authorizes appropriations.

10 & 11. Seng Mountain National Scenic Area and Bear Creek National Scenic Area (VA)—P.L. 111-11, Omnibus Public Land Management Act of 2009, Title I, Subtitle B—Virginia Ridge and Valley Wilderness, §1104; 2 pages.

- (a)(1)—designates 5,192-acre area as Seng Mountain National Scenic Area.
- (a)(2)—designates 5,128-acre area as Bear Creek National Scenic Area.
- (b)—establishes purposes:
  - "(1) to ensure the protection and preservation of scenic quality, water quality, natural characteristics, and water resources;"
  - "(2) to protect wildlife and fish habitat;"
  - "(3) to protect areas ... that may develop characteristics of old-growth forests; and"
  - "(4) to provide a variety of recreation opportunities."
- (c)(1)(B)—directs management in accordance with "the laws (including regulations) generally applicable to the National Forest System."
- (c)(2)—limits uses to those that "will further the purposes."
- (d)(1)—requires a management plan for the areas, as an amendment to the NFMA plan for the Jefferson National Forest, within two years.
- (e)—prohibits new roads, but "nothing ... denies any owner of private land (or an interest in private land) ... the right to access the private land."
- (f)—prohibits timber harvesting, except to "(A) control fire; (B) provide for public safety or trail access; or (C) control insect and disease outbreaks." Allows firewood collection for personal use along perimeter roads.
- (g)—authorizes insect and disease control "(1) to maintain scenic quality; (2) to prevent tree mortality; (3) to reduce hazards to visitors; or (4) to protect private land."
- (h)—allows "vegetation manipulation practices ... to maintain the visual quality and wildlife clearings."
- (i)—prohibits motorized vehicles, except "(A) to carry out administrative activities; (B) to assist [existing] wildlife management projects; and (C) during deer and bear hunting seasons" on specified roads.

- (k)—directs management that "maintains and enhances water quality."
- (l)—withdraws areas from mining and mineral leasing laws, subject to valid existing rights.

Wildlife Preserves

Congress has designated six wildlife preserves within the National Forest System.[34] These designations contrast with the other designations. First, they are much older, all having been designated prior to 1934; the oldest of all the other congressionally designated areas is Spruce Knob-Seneca Rocks NRA, designated in 1965. Also, the general management guidance for these areas is much less specific than for other congressionally designated areas, focusing exclusively on wildlife management activities. Finally, the preserves are more difficult to identify on the ground; none have reference maps showing the areas, and none identify the acreage encompassed. According to USFS data, the congressionally designated preserves range from less than 10,000 acres (for several units of the Ouachita and Ozark Preserves) to 612,736 acres for the Grand Canyon Preserve.[35]

1. Grand Canyon Preserve (AZ)—Act of June 29, 1906, ch. 3593; 1/2 page in Stats.

- §1—authorizes designation of an area to be "set aside for the protection of game animals and be recognized as a breeding place therefor."
- §2—prohibits "hunting, trapping, killing, or capturing of game animals, except under such regulations as may be prescribed from time to time by the Secretary."
- §3—establishes purposes: "to protect from trespass the public lands ... and the game animals which may be thereon, and not to interfere with the operation of the local game laws."

2. Norbeck Wildlife Preserve (SD)—Act of June 5, 1920, ch. 247; 1 page in Stats. Originally named Custer State Park Game Sanctuary.

- §1—authorizes designation an area not to exceed 30,000 acres "for the protection of game animals and birds and be recognized as a breeding place therefor."
- §2—prohibits "hunting, trapping, killing, or capturing of game animals and birds, except under such regulations as may be prescribed from time to time by the Secretary."
- §3—establishes purposes: "to protect from trespass the public lands ... and the game animals and birds which may be thereon, and not to interfere with the operation of the local game laws."
- §4—authorizes state to "erect and maintain a good substantial fence ... inclosing in whole or in part such areas" with gates for access for federal administration. "The right of the State to maintain this fence shall continue so long as the area ... is also given similar protection by the laws of the State of South Dakota."
- §5—authorizes a land exchange with the state.

3. Ozark Wildlife Preserve (AR; 5 units)—Act of Feb. 28, 1925, ch. 376; 1/2 page in Stats.

- §1—authorizes designation of lands to "be set aside for the protection of game animals, birds, or fish; and whoever shall hunt, catch, trap, willfully disturb, or kill any kind of game animal, game or non-game, bird, or fish, or take the eggs ... except under such general rules and regulations ... shall be fined ... or imprisoned ... or both: *Provided*, That no lands within the present limits of the fourth congressional district shall be included." (emphasis in original)

4. Tahquitz National Game Preserve (CA)—Act of July 3, 1926, ch. 776; 1 page in Stats.

- §1—creates the area "for the protection of game animals, and as the recognized breeding place therefor."
- §2—directs that "nothing ... shall prevent ... permitting other uses of said lands under and in conformity with the laws and rules and regulations applicable thereto so far as any such use may be consistent with the purposes."
- §3—prohibits "hunting, pursuing, poisoning, killing, or capturing by trapping, netting, or any other means, or attempting to hunt, pursue,

kill, or capture any wild animals or birds for any purpose ... except as hereinafter provided."

- §4—authorizes "needful rules and regulations for the administration ... , including regulations for hunting, capturing, or killing predatory animals, such as wolves, coyotes, cougar, or other species destructive to livestock or wild life."
- §5—authorizes land exchanges and land donations.

5. Ocala National Game Refuge (FL)—Act of June 28, 1930, ch. 709; 1/2 page in Stats.

- §1—authorizes designation of lands to be "set aside for the protection of game animals and birds, but it is not intended that the lands ... shall cease to be parts of the national forest ... and shall not prevent ... permitting other uses under and in conformity with the laws and regulation applicable thereto so far as such uses may be consistent with the purposes."
- §2—prohibits "hunting, pursuing, poisoning, killing, or capturing by trapping, netting, or any other means, or attempting to hunt, pursue, kill, or capture any game animals or birds ... except as herein provided."

6. Ouachita Wildlife Preserve (AR; 4 units)—Act of June 13, 1933, ch. 63;1/2 page in Stats.

- §1—authorizes presidential proclamation of lands for "providing breeding places and for the protection and administration of game animals, birds, and fish."
- §2—authorizes general rules and regulations.

Protection Areas

Congress has designated 38 areas within the National Forest System for some form of special protection. Each has distinct purposes and management direction enacted in the law designating the area. A few general designations are used—special management area, protection area, etc.— but many are unique and some areas have no title or name in the enabling legislation. Because of the broad array of purposes and guidance, no general conclusions can be drawn about these designations, or their differences compared to national monuments, recreation areas, and/or scenic areas. These designations are generally smaller than the monuments and recreation areas, but have a wide range of acreage, from less than 300 acres to 222,000 acres.

1. Special Management Unit adjacent to Washakie Wilderness (WY)[36]—P.L. 92-476, §5; ½ page.

- (a)—prohibits "harvesting of timber or public or private vehicular use of any existing road, and shall not construct or permit the construction or expansion of any road." Directs management "in accordance with the laws, rules, and regulations relating to the national forests especially to provide for nonvehicular access recreation and may construct such facilities and take such measures as are necessary for the health and safety of visitors and to protect the resources." Directs no effect on "such vehicular use and maintenance of existing roads as may be necessary for the administration."
- (b)—directs "a continuing study" and after five years, "shall recommend ... the area's highest and best public use."

2. Bull Run Watershed Management Unit (OR)—P.L. 95-200; 3+ pages.

- Preamble—contains findings.
- §1—designates 95,382-acre area, "subject to valid existing rights."
- §2(a)—directs management "as a watershed ... in accordance with the laws, rules, and regulations applicable to National Forest System lands ... except to the extent that any management plan or practice is found ... to have a significant adverse effect on compliance with water quality standards ... or on the quantity of water produced ... (and the Secretary shall take into consideration the cumulative effect of individually insignificant degradations), in which case ... the management plan and all relevant leases, permits, contracts, rights-of-way, or other rights or authorizations ... shall forthwith be altered ... to eliminate such adverse effect by application of different techniques or prohibitions of one or more such practices or uses: *Provided, however*, That use of such water for the production of energy and the transmission of such energy ... are deemed consistent with the purposes" and Bonneville Power Administration (BPA) rights-of-way

"are validated and confirmed and deemed consistent with the purposes." (emphasis in original)

- §2(b)—provides such management "through the development, maintenance, and periodic revision of land management [NFMA] plans, through the maintenance of systems for monitoring and evaluating water quality, and through supporting scientific research.... Provide for public participation and ... consult and coordinate with appropriate officials and advisors of the city, and shall consider such data and research as the city may collect.... Plans ... shall contain water quality standards ... which ... shall be substantially based on and shall reflect a quality of water not significant less than the quality reflected" in historic data.
- §2(c)—requires annual meeting with city officials to review impacts of proposed activities on water quantity and quality.
- §2(d)—directs that any disagreement between city and USFS over water quality standards or effects of activities go to an arbitration board.
- §2(e)—authorizes "regulations for controlling entry," including for city and agency officials, employees, contractors, and members of advisory groups, but has no effect on access to nonfederal lands or for authorized occupancies, except during fire or other emergencies.
- §3(a)—has no effect on "any lease, permit, contract, patent, right-of-way, or other land use right or authorization existing."
- §3(b)—has no effect on "any law governing appropriation or use of, or Federal right to, water on National Forest System lands; or as expanding or diminishing Federal, State, or local jurisdiction, responsibility, interests, or rights in water resources development or control."
- §3(d)—directs that "this Act shall take precedence over and supersede all State and local laws."
- §3(e)—allows challenges only "upon a showing of arbitrary, unreasonable, capricious, or illegal action or an absence of substantial good faith compliance."

3. Boundary Waters Canoe Area (BWCA) Mining Protection Zone (MN)—P.L. 95-495, Boundary Waters Canoe Area Wilderness Act; 21/2 pages.

- §1—establishes purposes: "for the protection, enhancement, and preservation of the natural values of the lakes, waterways, and associated forested areas ... and for the orderly management of public use and enjoyment of that area, while at the same time protecting the special qualities of the area as a natural forest-lakeland wilderness ecosystem of major esthetic, cultural, scientific, recreational and educational value."
- §9—designates 222,000-acre area "to protect existing natural values and high standards of environmental quality from the adverse impacts associated with mineral development."
- §11(a)—allows "no permit, lease, or other authorization ... for—(1) exploration for, or mining of, minerals owned by the United States; (2) exploration for, or mining of minerals within such areas if such activities may affect navigable waters; or (3) the use of property owned by the United States in relation to any mining of or exploration for minerals ... which may materially impair the natural values and environmental quality of the mining protection area." Applicable to BWCA wilderness and BWCA mining protection zone except as provided in the statute pursuant to a declared national emergency.
- §11(b)(1)—allows that "no permit, lease or other authorization shall be issued unless and until—"
    - "(A) the Secretary shall have approved a plan that details how mining will be conducted ... and ... how the area will be restored to its original condition or to a substantially equivalent condition, including the estimated cost thereof;"
    - "(B) the applicant has posted a bond for performance ... sufficient to assure completion of the reclamation plan;"
    - "(C) the applicant shall have obtained all permits, licenses, certification, and approvals required by Federal, State, or local law; and (iv) the Secretary has determined that no permanent facility will be constructed nor alteration will occur that could render the area incapable of reverting to its original condition or to a substantially equivalent condition."
- (c)—authorizes federal acquisition of minerals or mineral rights.
- (d)—provides guidance on legal action related to minerals or mineral rights.
- (e)—authorizes acquisition of "existing mineral interests by donation, purchase, exchange, or through exercise of the power of eminent domain."

- (f)—authorizes appropriations.

4. Area adjacent to Lee Metcalf Wilderness (MT)—P.L. 96-140, Lee Metcalf Wilderness Act of 1978, §2(c); 1/2 page.

- —presents findings “that certain lands ... near Monument Mountain have important recreational and wildlife values, including grizzly bear and elk habitat.”
- —designates 38,000-acre area, “managed to protect the wildlife and recreational values.”
- —withdraws area from mining and mineral leasing laws.
- —directs management “to maintain presently existing wilderness character, with no commercial timber harvest nor additional road construction permitted.”
- —allows “continued use ... by motorized equipment only for activities associated with existing levels of livestock grazing, administrative purposes (including snowmobile trail maintenance) and for snowmobiling during periods of adequate snow cover but only where such uses are compatible with the protection and propagation of wildlife.”
- —allows “limited motor vehicle access by individuals and others within the area where such access is compatible with the protection and propagation of wildlife and where such access was establish.”

*5. **Antone Bench Area*** (UT)—P.L. 98-428, Utah Wilderness Act of 1984, §306; 1 page.

- (a)—designates lands depicted on specified map. Directs management “subject to valid existing rights and until Congress determines otherwise.”
  - (1) withdraws area from mining and mineral leasing, except allows competitive leases for carbon dioxide (CO_2) for five years;
  - (2) allows 10-year CO_2 leases, and longer if commercial production occurs;
  - (3) allows “exploration ... only by helicopter or other methods which do not involve road construction or other significant surface disturbance.”
- (b)—limits development to include (1) minimum necessary road standards; (2) infrastructure camouflaged and located to minimize

intrusion; (3) mineral materials (e.g., sand and gravel) from outside wilderness; (4) seasonal construction to minimize recreation and wildlife impacts; (5) roads only for CO_2 extraction, closed to other vehicle access; (6) roads removed and reclaimed when mining operations cease; (7) waste disposed outside wilderness; and (8) no significant impairment of water quality or quantity.

6. Klamath River Basin Conservation Area (CA)—P.L. 99-552; 61/2 pages.

- §1—presents findings.
- §2(a)—directs Secretary to designate area.
- §2(b)—establishes program to:
  - 2(A)—"monitor and coordinate research ... [on] anadromous fish populations ... and evaluate the success of [restoration] activities."
  - 2(B)—directs actions to:
  - "(i) improve and restore ... habitats, and to promote access to blocked ... habitats;"
  - "(ii) rehabilitate problem watersheds;"
  - "(iii) improve existing ... hatcheries and rearing ponds'"
  - "(iv) implement an intensive, short-term stocking program; and"
  - "(v) improve upstream and downstream migration by removal of obstacles to fish passage."
- §3—establishes Klamath Fishery Management Council; defines functions, membership, procedures [2 pages].
- §4—establishes Klamath River Basin Fisheries Task Force; defines functions, membership [11/2 pages].
- §5—requires an MOU to facilitate fisheries regulations enforcement.
- §6—authorizes appropriations and establishes cost-sharing standards.

*7. **Kings River Special Management Area*** (CA)—P.L. 100-150, §2; 2 pages.

- (a)—designates area "to provide for public outdoor recreation use and enjoyment, to protect those areas' natural, archaeological, and scenic resources, and to provide for appropriate fish and wildlife management."
- (c)—directs management "in accordance with ... the provisions of law generally applicable to units of the National Forest System." Has no

effect on grazing; allows "the cutting of timber ... only in those cases where ... required ... to control the attacks of fire, insects, or diseases or to otherwise conserve the scenery or the natural or historical objects."

- (d)—withdraws area from mining and mineral leasing laws, subject to valid existing rights.
- (e)—permits hunting and fishing. Allows zones and periods of no hunting or fishing for public safety, administration, fish and wildlife management or public use and enjoyment; requires consultation with state.
- (f)—requires a management plan within three years to: "provide for public outdoor recreation use and enjoyment, protect the area's natural, archaeological, and scenic resources, and provide for appropriate fish and wildlife management. [It] shall contain provisions for management of vegetation ... designed to enhance the wildlife carrying capacity, permit off-road vehicular use of off-road trails to the same extent and in the same locations as was permitted before, [and] provide for the development of hiking trails."
- (g)—directs the Secretary to "provide the State or private owner, claimant, or occupier and their successors in interest such rights as may be necessary to assure adequate and feasible access for economic and other purposes ... subject to reasonable regulations ... to protect the natural and other values ... taking into account the traditional and customary means of access."
- (h)—directs that "no Federal lands may be used for the construction of any dam or diversion."

8 & 9. Beech Creek Botanical Area and Robert S. Kerr Arboretum, Natural Center, and Botanical Area (OK)—P.L. 100-499, Winding Stair Mountain National Recreation Area and Wilderness Act, §9; 1 page.

- (a)—establishes purposes: "to protect and interpret to the public area ... which contain unique plant species and unique plant communities that are significant in their occurrence, variety and location."
  - (1)—designates 8,026-acre Robert S. Kerr Arboretum, Natural Center, and Botanical Area.
  - (2)—designates 400-acre Beech Creek Botanical Area.
- (c)—directs management "in accordance with the laws, rules and regulations applicable to the national forests.... Except as provided in

section 16 of this Act, vegetative manipulation, including the cutting of trees, shall be permitted in such areas only when necessary for the protection and interpretation of the unique plant species and unique plant communities." Allows "expansion of roads, improvements and other facilities" in the Robert S. Kerr Arboretum, Natural Center, and Botanical Area.

- §16—allows "measures ... necessary in the event of fire, or infestation of insects or disease or for public health and safety."

11-22. Tongass National Forest Land Use Designation (LUD) II Management Areas (AK)—P.L. 101-626, Tongass Reform Act, §201; 1 page.

- Designates 12 areas with 722,482 total acres: Yakutat Forelands; Berners Bay; Anan Creek; Kadashan; Lisianski River/Upper Hoonah Sound; Mt. Calder/Mt. Holbrook; Nutkwa; Outside Islands; Trap Bay; Point Adolphus/Mud Bay; Naha; and Salmon Bay. Specifies that the areas "are hereby allocated to Land Use Designation II ('LUD II') as described in the Tongass National Forest Land Management Plan, completed March 1979, and amended Winter 1985-1986, and shall be managed ... in perpetuity in accordance with such designation."

23 & 24. East Fork and Eagle Creek Areas (IL)—P.L. 101-633, Illinois Wilderness Act of 1990, §10; 1/2 page.

- (a)(1)—allows mining and prospecting for fluorspar and associated minerals within the areas. Directs that "these lands shall also be managed, to the extent practicable, to preserve their potential for future inclusion in the National Wilderness Preservation System."
- (a)(2)(A)—designates 2,042-acre East Fork Area.
- (a)(2)(B)—designates 722-acre Eagle Creek Area.
- (b)—allows "prospecting for fluorspar and associated minerals ... for a period of not more than 8 years.... If significant deposits ... are found, then mining for those minerals may be allowed for a 20-year period."
- (c)—has no effect on mining and prospecting permits and rights.
- (d)—directs that, after expiration of periods in (b), the lands "shall be designated as wilderness."

25. Greer Spring Special Management Area (MO)—P.L. 102-220, Greer Spring Acquisition and Protection Act of 1991, §4; 1 page.

- (a)—designates area "to provide for public outdoor recreation use, including fishing and hunting, in a natural setting, and the enjoyment, to protect those areas' natural, archaeological, and scenic resources, and to provide for appropriate resource management" and directs management "in accordance with ... provisions of law generally applicable to units of the National Forest System."
- (c)—permits "the harvesting of timber ... only in those cases where ... required ... to control insects or disease, for public safety, for salvage sales, or to accomplish the objectives.... To the extent practicable, timber harvesting shall be conducted only by the individual tree selection method."
- (d)—permits hunting and fishing.
- (e)—withdraws area from mining and mineral leasing laws, subject to valid existing rights.
- (f)—limits road construction and maintenance to "only those roads ... which are indicated on the map: *Provided*, that the Secretary shall provide access to such roads, or to timber harvesting, in such a manner as to minimize environmental impact." (emphasis in original)
- §5—authorizes appropriations.

26. Bowen Gulch Protection Area (CO)—P.L. 103-77, Colorado Wilderness Act of 1993, §6; ½ page.

- (a)—designates 11,600-acre area.
- (b)—directs management "in accordance with ... the laws and regulations generally applicable to the National Forest System."
- (c)—withdraws area from mining and mineral leasing, subject to valid existing rights.
- (d)—prohibits developed campgrounds and new roads or trails.
- (e)—prohibits timber harvesting "except to the extent that would be permitted in wilderness ... for necessary control of fire, insects, and diseases, and for public safety."
- (f)—allows motorized travel "only on those designated trails and routes existing as of July 1, 1991 and only during periods of adequate snow cover. At all other times, mechanized, non-motorized travel shall be permitted."

- (g)—requires a management plan for the area during revision of the NFMA plan for the Arapaho National Forest.

27, 28, & 29. Piedra, Roubideau, and Tabeguache Areas (CO)—P.L. 103-77, Colorado Wilderness Act of 1993, §9; 1 page.

- (a)—designates:
  - (1)—62,550-acre Piedra Area.
  - (2)—19,650-acre Roubideau Area.
  - (3)—17,240-acre Tabeguache Area.
- (b)(1)—withdraws areas from mining and mineral leasing laws, subject to valid existing rights.
- (b)(3)—directs management "until Congress determines otherwise, ... so as to maintain the areas' presently existing wilderness character and potential for inclusion in the National Wilderness Preservation System."
- (b)(4)—has no effect on livestock grazing, except for the ban on mechanized or motorized travel.
- (c)—directs compilation of "data concerning the water resources ... and existing and proposed water resource facilities."

30. Kelly Butte Special Management Area (WA)—P.L. 105-277, Omnibus Consolidated and Emergency Supplemental Appropriations, 1999, Div. A, §101(e), Title VI, §611; 1/2 page.

- (a)—designates 5,642-acre area, upon conveyance of the land to the United States.
- (b)—directs management "in accordance with the laws, rules and regulations generally applicable to National Forest System lands ... (1) with special emphasis on:"
  - "(A) preserving its natural character and protecting and enhancing water quality;"
  - "(B) permitting hunting and fishing;"
  - "(C) providing opportunities for primitive and semi-primitive recreation and scientific research and study;"
  - "(D) protecting and enhancing populations of fish, wildlife and native plant species; and"
  - "(E) allowing for traditional uses by native American peoples;"
- (b)(2)—prohibits "commercial timber harvest and road construction."

- (b)(3)—prohibits access by "motor vehicles, except as may be necessary for administrative purposes or in emergencies (including rescue operations) to protect public health and safety."
- (b)(4)—withdraws area from mining and mineral leasing laws, subject to valid existing rights.
- (c)—prohibits buffer zones.

31. Granite Watershed Area (CA)—P.L. 105-281, Granite Watershed Enhancement and Protection Act of 1998; 2 pages.

- §2(a)—authorizes "a contract with a single private contractor to perform multiple resource management activities ... for ... demonstrating enhanced ecosystem health and water quality, and significant reducing the risk of catastrophic wildfire.... The contract shall be for a term of 5 years."
- §2(b)—authorizes activities for:
  - "(1) Reduction of forest fuel loads through the use of precommercial and commercial thinning and prescribed burns."
  - "(2) Monitoring of ecosystem health and water quality."
  - "(3) Monitoring of the presence of wildlife ... and the effect of the activities on wildlife presence."
  - "(4) Such other resource management activities as the Secretary considers appropriate to demonstrate enhanced ecosystem health and water quality."
- §2(c)—requires activities to comply with federal law and "the standards and guidelines for the conservation of the California spotted owl."
- §2(d) & (e)—authorizes appropriations from several accounts, offsets from forest products removed, and acceptance of state funds.
- §2(f)—requires an annual report to Congress.

32. James Peak Protection Area (CO)—P.L. 107-216, James Peak Wilderness and Protection Area Act, §3; 3 pages

- (a)(1)—presents findings.
- (a)(2)—identifies purpose: "to provide management of certain lands ... in a manner consistent with the 1997 Revised Land and Resources Management Plan ... to protect the natural qualities."
- (b)—designates 16,000-acre area.

- (d)(1)—directs management, in general, as in the 1997 revision of the NFMA plan for the Arapaho/Roosevelt National Forest and Pawnee National Grassland. Also:
    - (A)—has no effect on grazing.
    - (B)—withdraws area from public land, mining, mineral leasing laws, subject to valid existing rights.
    - (C)—requires a road and trail inventory; allows connections among and closures of roads and trails; and restricts "motorized and mechanized travel ... only on those roads and trails identified as open to use."
    - (D)—prohibits "new roads or trails ... except: (i) ... to replace roads or trails ... which have become nonserviceable through reasons other than neglect. (ii) Nonpermanent roads as needed for hazardous fuels reduction or other control of fire, insect or disease control projects, or other management purposes." (iii) for private land access, or (iv) & (v) for 2 specific trails.
    - (E)—prohibits "timber harvesting ... except to the extent needed for hazardous fuels reduction or other control of fire, insect or disease control projects, of protection of public health or safety."
- (d)(2)—allows "maintenance of rights-of-way and access roads ... to operate the natural gas pipeline.... Such maintenance may including vegetation management, road maintenance, ground stabilization, and motorized vehicle access."
- (e)(1)—directs that "nothing shall constitute ... either an express or implied reservation of any water or water rights ... [or] affect any conditional or absolute water rights."
- (e)(3)—directs that "nothing ... shall affect, impede, interfere with, or diminish the operation, existence, access, maintenance, improvement, or construction of water facilities and infrastructure, rights-of-way, or other water-related property, interests, and uses."

33. T'uf Shur Bien Preservation Trust Area (NM)—P.L. 108-7, Consolidated Appropriations Resolution, 2003, Division F, Title IV—T'uf Shur Bien Preservation Trust Area Act; 15 pages.

- §402(a)—presents findings.
- §402(b)—identifies purposes: (1) to establish the trust area; (2) to settle lawsuit; and (3) to provide a "fair and just settlement of the Pueblo's claim."

- §403—includes definitions.
- §404(a)—designates 9,890-acre area.
- §404(b)(1)—directs management "as part of the National Forest System."
- §404(b)(2)—prohibits limits on Native uses, except under the Wilderness Act and federal wildlife law.
- §404(e)—prohibits certain uses: (1)(B)(i) gaming or gambling; (ii) mineral production; (iii) timber production; (iv) any new use objected to by Pueblo; (2) "closed to the location of mining claims."
- §405—identifies Pueblo rights and interests: (a) "(1) Free and unrestricted access ... for traditional or cultural uses.... (2) Perpetual preservation of the national forest and wilderness character ..."
- §406—identifies limitations on Pueblo rights and interests: no right to "(1) ... sell, grant, lease, convey, encumber, or exchange land or any interest in land; (2) any exemption from applicable Federal wildlife protection laws."
- §407—directs consultation "with the Pueblo not less than twice each year ... concerning protection, preservation, and management ... (including proposed new uses and modified uses ... and authorizations ...)." Requires public notice of any proposed new or modified use.
- §408—defines legal jurisdictions: (a) criminal jurisdictions of federal, Pueblo, and state governments; (b) civil jurisdiction of Pueblo; (c) prohibits taxation.
- §409—generally excludes subdivisions from area and from Pueblo jurisdiction.
- §410—extinguishes claims.
- §411—establishes certain rules of construction: "(a) ... recognizes only enumerated rights and interests; (b) ... valid private property rights ... are not modified or otherwise affected; (c) not construed as precedent for ... management of the" NFS; (d) has no effect on state wildlife responsibilities.
- §412—defines judicial review: enforcement, waiver, venue.
- §413—allows contributions and land exchanges.
- §414—authorizes appropriations.

34. Crystal Springs Watershed Special Resources Management Unit (OR)—P.L. 111-11, Omnibus Public Land Management Act of 2009, Title I, Subtitle C—Mt. Hood Wilderness, Oregon, §1205(a); 21/2 pages.

- (1)(A)—establishes area upon completion of land exchange.
- (1)(C)—withdraws area from public land, mining, and mineral leasing laws, subject to valid existing rights.
- (2)—identifies purposes:
  - "(A) to ensure the protection of the quality and quantity of the ... watershed as a clean drinking water source."
  - "(B) to allow visitors to enjoy the special scenic, natural, cultural, and wildlife values."
- (4)(A)—directs management "in accordance with the laws (including regulations) and rules applicable to units of the National Forest System ; and ... only allow uses ... consistent with the purposes."
- (4)(B)—allows, "to protect the water quality, water quantity, and scenic, cultural, natural, and wildlife values, fuel reduction and forest health treatments to maintain and restore fire-resilient forest structures containing late successional forest structure characterized by large trees and multistoried canopies, as ecologically appropriate" within 400 feet of structures or certain roads, and on NFS lands "with priority given ... to restore previously harvested areas, including the removal of logging slash, smaller diameter material, and ladder fuels."
- (5)—prohibits, subject to valid existing rights:
  - (A) new road construction or renovation of existing roads, "except as necessary to protect public health and safety."
  - (B) commercial timber harvesting, except as byproducts of activities under §1205(a)(4)(B).
  - (C) commercial livestock grazing.
  - (D) placement of new fuel storage tanks.
  - (E) "the application of any toxic chemicals (other than fire retardants), including pesticides, rodenticides, or herbicides."
- (6)—allows road closures, except for one specified road.
- (7)—has no effect on "the use of, or access to, any private property" by the owners or their guests.
- (8)—authorizes land acquisition.

35 & 36. Upper Big Bottom and Cultus Creek Areas (OR)—P.L. 111-11, Omnibus Public Land Management Act of 2009, Title I, Subtitle C—Mt. Hood Wilderness, Oregon, §1205(b); 1 page.

- (1)—directs management that "preserves the natural and primitive character of the land for recreational, scenic, and scientific use."

- (2)(A)—designates 1,580-acre Upper Big Bottom.
- (2)(B)—designates 280-acre Cultus Creek.
- (4)(A)—allows "only uses ... consistent with the purposes," subject to valid existing rights.
- (4)(B)—prohibits: (i) permanent roads; (ii) commercial enterprises; and (iii) motor vehicles and temporary roads, "except as necessary to meet minimum requirements for the administration ... and to protect public health and safety."
- (5)—withdraws the areas from public land, mining, and mineral leasing laws, subject to valid existing rights.

37. Ancient Bristlecone Pine Forest (CA)—P.L. 111-11, Omnibus Public Land Management Act of 2009, Title I, Subtitle K—Eastern Sierra and Northern San Gabriel Wilderness, California, §1808; 1 page.

- (a)—designated 31,700-acre area "to conserve and protect the Ancient Bristlecone Pines by maintaining near-natural conditions and to ensure the survival of the Pines for ... public enjoyment and scientific study."
- (c)(1)—directs management:
  - (A) to: "(i) protect the resources and values of the area; and (ii) promote:"
    - "(I) the protection of bristlecone pines for public enjoyment and scientific study;"
    - "(II) the recognition of the botanical, scenic, and historical values; and"
    - "(III) the maintenance of near-natural conditions by ensuring that all activities are subordinate to the needs of protecting and preserving bristlecone pines and wood remnants; and"
  - "(B) "in accordance with" NFMA "and other applicable laws."
- (c)(3)—withdraws area from public land, mining, mineral leasing, and mineral materials laws, subject to valid existing rights.

38. Wyoming Range Withdrawal Area (WY)—P.L. 111-11, Omnibus Public Land Management Act of 2009, Title III, Subtitle C—Wyoming Range; 2 pages.

- §3201—provides definitions.

- §3202(a)—withdraws area from public land, mining, and mineral leasing laws, subject to valid existing rights.
- §3202(b)—authorizes withdrawal of land from which valid existing rights are acquired under §3203.
- §3202(c)—prohibits buffer zone.
- §3202(d)—directs management consistent with the NFMA plan for the BridgerTeton National Forest.
- §3202(e)—has no effect on actions associated with prior lease sales.
- §3202(f)—authorizes oil and gas leases within 1 mile inside the boundary, provided that:
    - "(1) The lease may only be accessed by direction drilling from" an existing lease, adjacent to and outside the area.
    - "(2) The lease shall prohibit, without exception or waiver, surface occupancy and surface disturbance for any activities."
    - "(3) The directional drilling may extend no further than 1 mile inside" the area.
- §3203(a)—directs notification of valid existing rights-holders.
- §3203(b)—authorizes "the retirement and repurchase of that right."
- §3203(c)—prohibits the use of federal funds to purchase such rights.
- §3203(d)—authorizes acceptance, and subsequent cancelation, of that right.
- §3203(e)—directs that this authority has no effect on any authority "to modify, suspend, or terminate a lease without compensation, or to recognize the transfer of a valid existing mining or leasing right, if otherwise authorized by law."

Other Designations

Congress has enacted 11 other designations that do not have common purposes or management schemes as described in the previous sections of this appendix. Two are historic sites—the Cradle of Forestry, which is largely the Biltmore Estate, where Gifford Pinchot (the first USFS chief) first practiced forestry; and Grey Towers, the Pinchot family estate. Another one is a special site for astronomical research. Four provide management direction that seems to differ little from the general management direction for National Forest System lands and units. One provides special mining access at the edge of a designated wilderness; another provides special allocation of timber revenues. The last two are unique—a tallgrass prairie and a working ranch/forest intended to be managed on a self-sustaining basis. These areas range in size from less than 5,000 acres to nearly 550,000 acres.

1. Cradle of Forestry in America (NC)—P.L. 90-398; 1 page.

- §1—designates the site "to preserve, develop, and make available to this and future generations the birthplace of forestry and forestry education in America and to promote, demonstrate, and stimulate interest in and knowledge of the management of forest lands under principles of multiple use and sustained yield and the development and progress of management of forest lands in America."
- §2—directs management "within and as a part of the Pisgah National Forest ... and for such management, utilization, and disposal of the natural resources as ... will promote or is compatible with and does not significant impair the purposes."
- §3—requires cooperation with "public and private agencies and organizations and individuals."

2. Alpine Lakes Area Management Unit (WA)—P.L. 94-357; 41/2 pages.

- §2(a)—presents findings.
- §2(b)—designates the area "to provide for public outdoor recreation and use and for economic utilization of commercial forest lands, geological features, lakes, streams and other resources" subject to valid existing rights.
- §4—authorizes land acquisition and directs a land exchange (11/2 pages).
- §6(a)—requires, "in accordance with applicable acts governing the administration of the National Forest system and with full public participation ... prepare, complete, and begin to implement ... a single multiple use plan" within two years.
- §7(a)—permits hunting and fishing. Requires consultation with state on any regulations. Has no effect on state jurisdiction or responsibilities.
- §7(b)—has no effect on state or local governmental "right to exercise civil and criminal jurisdiction ... or its right to tax persons, corporations, franchises, or other non-Federal property."
- §8—authorizes appropriations.

3 & 4. Gospel-Hump Management Areas and Development Areas (ID)—P.L. 95-237, Endangered American Wilderness Act of 1978, §4; 21/2 pages.

- (a)(2)—directs that management areas "shall be managed in accordance with the multipurpose resource development plan." Designates 92,000 acres.
- (a)(3)—directs that development areas "shall be immediately available for resource utilization under the existing applicable Forest Service land management plans." Designates 45,000 acres.
- (b)(1)—establishes advisory committee; defines membership, procedures.
- (c)(1)—requires cooperation with state and Department of the Interior on "comprehensive fish and game research program within the Gospel-Hump Area and surrounding Federal lands" including "resident and anadromous fisheries resources (including water quality relationships) and the status, distribution, movements, and management of game populations" for "integration of land management and development with the protection and enhancement of these fish and game resources."
- (d)(1)—requires, within four years, implementation of a multipurpose resource development plan for the management areas.
- (d)(2)—requires plan to be consistent with MUSYA and NFMA, including "field data on soil types and soil hazards, and to consider timber volumes, timber site classes, and productivity."
- (d)(3)—requires "adequate public involvement, and ... full use of the recommendations" of the advisory committee.
- (d)(4)—requires annual review of plan.
- (f)—includes timber resources of development areas and management areas in determining allowable sale level in the NFMA plan for the Nez Perce National Forest.
- (g)—directs that "nothing ... shall prevent ... any activity, including prospecting, for the purpose of gathering information about mineral or other resources, if such activity is carried on in a manner compatible with the preservation of the wilderness environment.... [Requires areas to] be surveyed on a planned recurring basis ... to determine the mineral values, if any, that may be present."
- (h)—authorizes appropriations.

5. Special Mining Management Zone—Clear Creek (ID)—P.L. 96-312, Central Idaho Wilderness Act of 1980, §5(d); 1 page.

- (1)—designates area within River of No Return Wilderness [now Frank Church-River of No Return Wilderness]. Provides that "all prospecting and exploration for, and development or mining of cobalt and associated minerals shall be considered a dominant use of such land." Directs management "subject to such laws and regulations as are generally applicable to National Forest System lands ... including such laws and regulations which relate to the right of access to valid mining claims and private property: *Provided*, That:"
    - "(A) all mining locations and associated access roads shall be held and used solely for mining or mineral processing operations and uses reasonably incident thereto, except that the Secretary may permit such access roads to be utilized by the State of Idaho to facilitate the management of bighorn sheep in the Special Management Zone;"
    - "(B) subject to valid existing rights, all patents ... within the Special Management Zone shall convey title to the mineral deposits" but not to the land. "The patentee shall have the right to cut and use as much mature timber therefrom as may be needed in the extraction, removal, and beneficiation of the mineral deposits, if such needed timber is not otherwise reasonably available, and if such timber is cut under sound principles of forest management ... [Also] the patentee shall have the right to use as much of the surface as reasonably necessary for the mining, removal, extraction, or beneficiation of the mineral deposits."
    - "(C) The Secretary may take all reasonable measures to see that the mining or processing of cobalt and associated minerals ... does not significantly impair the overall habitat of the bighorn sheep."
- (2)—requires, within three years, Secretary of Defense to report on "the strategic significance of the materials and minerals found in" the Zone.

6. Langmuir Research Site (NM)—P.L. 96-550, New Mexico Wilderness Act of 1980, Title II; 1 page.

- §201—designates area "to encourage scientific research into atmospheric processes and astronomical phenomena, and to preserve conditions necessary for that research."
- §202—presents finding.
- 203—directs management "in accordance with the laws, rules, and regulations applicable to National Forest System lands."
- §204—requires a land use agreement with the New Mexico Institute of Mining and Technology to establish conditions of land use and working relationships.
- §205(a)—designates area of 30,000 acres, with 1,000-acre principle research facility.
- §205(b)—requires, as soon as practicable, a comprehensive management plan, incorporated into the NFMA plan for the Cibola National Forest.
- §205(d)—requires consultation on plan with National Science Foundation, NM Institute of Mining and Technology, NM Academy of Sciences "and appropriate conservation, wilderness, wildlife, industry, and other public interest groups."

7. Quinault Special Management Area (WA)—P.L. 100-638; 3+ pages.

- §2—designates 5,460-acre area.
- §3(a)—transfers some USFS lands to DOI, in trust for the tribe.
- §3(b)—directs management in "accordance with the laws, rules and regulations applicable to the national forests."
- §3(c)—directs that tribal rights to revenues from area "shall not affect the management of these lands nor create a trust or fiduciary duty."
- §4(a)—directs that the USFS retains 10% of "gross proceeds from the sale of forest products."
- §4(b)(1)—directs that 45% of receipts are deposited in a special account [see §8].
- §4(b)(2)—directs that 45% of receipts are for USFS payments to states for roads and schools in the county.
- §5(a)—prohibits sales for overseas export of unprocessed timber.
- §5(b)(1)—limits sale level to the non-declining even-flow of timber harvest, with departure from that level for multiple-use objectives, and annual fluctuations around decade average.[37]
- §5(b)(2)—directs administration "in accordance with the conditions of the Policy Statement for the Grays Harbor sustained yield unit."

- §6—requires "permanent easements for ... continuing access, including public access, to" NFS lands.
- §7—allows additional rights-of-way for land management.
- §8—directs that the funds in the special account [under §4(b)(1)] be used for (a) Indian timber sale and associated road costs; (b) mitigation of adverse environmental impact; (c) reforestation; or (d) purchase of lands or interests from willing non-Indian sellers.

8. Midewin National Tallgrass Prairie (IL)—P.L. 104-106, National Defense Authorization Act for FY1996, Title XXIX—Illinois Land Conservation Act of 1995, Subtitle A; 101/2 pages.

- §2902—provides definitions, including MNP as the area.
- §2911—identifies principles of transfer.
- §2912—transfers management responsibilities and jurisdiction from DOD to USDA [21/2 pages].
- §2913—provides continued DOD responsibility and liability for hazardous materials.
- §2914(a)—establishes area.
- §2914(b)(1)—directs management "as a part of the National Forest System in accordance with ... the laws, rules, and regulations pertaining to the National Forest System."
- §2914(b)(3)—directs, in developing a land and resource management plan, consultation with the Illinois Department of Natural Resources and local governments.
- §2914(c)—established purposes:
    - "(1) To manage the land and water resources of the MNP in a manner that will conserve and enhance the native populations and habitats of fish, wildlife, and plants."
    - "(2) To provide opportunities for scientific, environmental, and land use education and research."
    - "(3) To allow the continuation of agricultural uses of lands."
    - "(4) To provide a variety of recreation opportunities that are not inconsistent with the preceding purposes."
- §2914(d)—authorizes land acquisition.
- §2914(e)—authorizes and encourages cooperation "with appropriate Federal, State, and local governmental agencies, private organizations and corporations" including through cooperative agreements. "The objects of such cooperation may include public education, land and

resource protection, and cooperative management among government, corporate, and private landowners."

- §2915(a)—prohibits any new highway or public road through or across any portion of the area. Allows "construction and maintenance of roads for use within the MNP, the granting of authorizations for utility rights-of-way, or such access as is necessary ... [including] access by the Secretary of the Army for purposes of restoration and cleanup."
- §2915(b)—allows special use authorizations for agricultural purposes, including to supplant DOD agricultural leases.
- §2915(c)—directs that receipts be included for USFS payments to states, and allows remaining receipts to be available for prairie improvement work.
- §2915(d)—authorizes "reasonable fees for the admission, occupancy, and use," with a schedule for reduced fees or waivers for volunteer services, research, and education.
- §2915(e)—authorizes the salvage sale of any transferred facilities and improvements.
- §2915(f)—directs deposits of fees and salvage receipts into a special fund "for restoration and administration of the ... [area], including construction of a visitor and education center, restoration of ecosystems, construction of recreational facilities (such as trails), construction of administrative offices, an operation and maintenance of the MNP."
- §2916—identifies special transfer rules for specific areas [2 pages].

9. Valles Caldera National Preserve (NM)—P.L. 106-248, Valles Caldera Preservation Act; 15+pages.

- §102(a)—presents findings.
- §102(b)—identifies purposes:
    - "(1) to authorize Federal acquisition of the Baca ranch;"
    - "(2) to protect and preserve for future generations the scientific, scenic, historic, and natural values ... including rivers and ecosystems and archaeological, geological, and cultural resources;"
    - "(3) to provide opportunities for public recreation;"

    - "(4) to establish a demonstration area for an experimental management regime adapted to this unique property ... to promote long term financial sustainability; and"
    - "(5) to provide for sustained yield management of Baca ranch for timber production and domesticated livestock grazing insofar as is consistent with the other purposes."
- §103—includes definitions.
- §104(a)—authorizes acquisition of Baca ranch—94,761 acres—at appraised fair market value.
- §104(d)—requires a watershed management report, within 120 days, on alternatives to coordinate land management and "allow for improved management of elk and other wildlife populations."
- §104(e)—makes acquisition "subject to valid existing mineral interests." Authorizes acquisition of subsurface estate.
- §105(a)—establishes Valles Caldera National Preserve as "a unit of the National Forest System."
- §105(b)—identifies purposes: "to protect and preserve the scientific, scenic, geologic, watershed, fish, wildlife, historic, cultural, and recreational values, and to provide for multiple use and sustained yield of renewable resources."
- §105(e)—(1) withdraws area from mineral leasing, subject to valid existing rights; (2) allows for "common varieties of mineral materials ... for ... roads and facilities within the Preserve."
- §105(f)—has no effect on state fish and wildlife responsibilities. Allows zones and periods of no hunting or fishing for "public safety, administration, the protection of nongame species and their habitats, or public use and enjoyment; requires consultation with state.
- §105(g)(1)—"For ... preserving the natural, cultural, religious, and historic resources on Redondo Peak" prohibits roads, structures, or facilities and motorized access above 10,000 feet.
- §105(g)(2)—allows exceptions for existing road and trail maintenance, for new trails and needed road relocation "to avoid Native American religious and cultural sites;" and for administration, including emergencies.
- §106—establishes Valles Caldera Trust (a) "a wholly owed government corporation (b)(1) to provide management and administration services; (2) to establish and implement management policies; (3) to receive and collect funds from private and public sources and to make dispositions in support of the management and

Administration of the Preserve; and (4) to cooperate with Federal, State, and local governmental units, and with Indian tribes and Pueblos." (d) authorizes staff; (e) establishes the Trust as a government corporation; (f) exempts the Trust from state and local taxes; (g) allows donations; (h) allows retention and use of receipts. [21/2 pages].

- §107—creates a Board of Trustees; identifies membership [2 pages].
- §108(a)—directs resource management by the Trust
- §108(b)—identifies management responsibilities as:
    - "(1) administration ... of the Preserve;"
    - "(2) preservation and development of the land and resources;"
    - "(3) interpretation of the Preserve and its history;"
    - "(4) management of public use and occupancy; and"
    - "(5) maintenance, rehabilitation, repair, and improvement of property within the Preserve."
- §108(d)—requires a comprehensive management program, within two years, for:
    - "(1) operation of the Preserve as a working ranch;"
    - "(2) the protection and preservation of the scientific, scenic, geologic,
      watershed, fish, wildlife, historic, cultural and recreational values;"
    - "(3) multiple use and sustained yield of renewable resources;"
    - "(4) public use of and access to ... for recreation;"
    - "(5) renewable resource utilization and management alternatives that—(A) benefit local communities and small businesses; (B) enhance coordination of management objectives with those on surrounding N.F.S. land; and (C) provide cost savings; and"
    - "(6) optimizing the generation of income based on existing market conditions, to the extent that it does not unreasonably diminish the long-term scenic and natural values ... of the ... capability of the land."
- §108(e)—authorizes public use and recreation, (1) "including, but not limited to camping and picnicking, hiking, and cross country skiing; (2) reasonable fees for admission to, and the use and occupancy of the Preserve; [and] (3) reasonable access ... [but] reasonably limit the number and types of recreational admissions ... based on the capability of the land, resources, and facilities."

- §108(f)—(1) directs management "in conformity with ... all laws pertaining to the National Forest System, except" NFMA; (2) deems the Trust "a Federal agency for purposes of compliance with Federal environmental laws;" (4) authorizes "a compilation of rules and regulations which ... are inappropriate, incompatible with this title, or unduly burdensome;" (5) directs cooperation with Indian tribes and Pueblos and authorizes religious and cultural uses; and (6) prohibits administrative appeals.
- §109(a)—authorizes the Secretary to: continue to certain management prerogatives: issue rights-of-way; implement prohibitions generally applicable to the NFS; implement Wild and Scenic Rivers Act and Federal Power Act authorities; acquire mineral rights; enforce laws and provide fire management; exchange land or interests; refer civil and criminal cases to Department of Justice; retain title to fossils and archaeological artifacts; build and operate a visitor center; and assess Trust performance.
- §109(b)—directs the Secretary to manage the area until the Trust is in place.
- §110—terminates the Trust after 20 years, with reviews and recommendations from the Board on goal achievement after 15 years and on extension or termination after18 years.
- §111—(a) authorizes appropriations; (b) requires planned "annual decreasing appropriated funds that will achieve, at a minimum, the financially self-sustained operation of the Trust within 15 full fiscal years."
- §112—requires GAO studies three and seven years after Trust management begins.

10. Grey Towers National Historic Site (PA)—P.L. 108-447, Consolidated Appropriations Act, 2005, Division E, §348; 3 pages.

- (b)(1)—presents findings.
- (b)(2)—identifies purposes:
    - "(A) To honor and perpetuate the memory of Gifford Pinchot."
    - "(B) To promote the recreational and educational resources of Milford, Pennsylvania, and its environs."
    - "(C) To authorize the Secretary of Agriculture—(i) to further the scientific, policy analysis, educational, and cultural programs in natural resource conservation at Grey Towers; (ii) to manage the

property and environs more efficiently and effectively; and (iii) to further collaborative ties with the Pinchot Institute for Conservation and other Federal, State, and local agencies with shared interests."

- (b)(3)—includes definitions.
- (c)—designates the site.
- (d)(1)—directs management for:
    - "(A) Education, public demonstration projects, and research related to natural resource conservation, protection, management, and use."
    - "(B) Leadership development within the natural resource professions and the Federal civil service."
    - "(C) Continuing Gifford Pinchot's legacy through pursuit of new ideas, strategies, and solutions to natural resource issues that include economic, ecological, and social values."
    - "(D) Preservation, use, and maintenance of the buildings, grounds, facilities, and archives associated with Gifford Pinchot."
    - "(E) Study and interpretation of the life and works of Gifford Pinchot."
    - "(F) Public recreation and enjoyment."
    - "(G) Protection and enjoyment of the scenic and natural environs."
- (d)(2)—directs management "as components of the National Forest System in accordance with ... laws generally applicable to the administration of national historic sites, and the laws, rules, and regulations applicable to the National Forest System, except" for NFMA.
- (d)(3) & (4)—authorize land acquisition and acceptance of donations.
- (e)—authorizes grants, contracts, and cooperative agreements, including interdepartmental cooperation with DOI.
- (f)—authorizes "reasonable fees and charges" and allows use of the receipts to support programs.

11. Pickerel Hill Management Area (CA)—P.L. 111-11, Omnibus Public Land Management Act of 2009, Title I, Subtitle K—Eastern Sierra and Northern San Gabriel Wilderness, California, §1807; 1/4 page.

- designates 3,690-acre area, to "be managed in a manner consistent with the non-Wilderness forest areas immediately surrounding, including the allowance of snowmobile use."

End Notes

1 The act also designated 33 new wilderness areas, added to 7 existing wilderness areas, established at least 25 wild and scenic rivers, and created three new conservation areas on lands administered by the Bureau of Land Management and the National Park Service, in the Department of the Interior.

2 For information on pending wilderness legislation (which also commonly includes other congressional land area designations), see CRS Report R40237, *Federal Lands Managed by the Bureau of Land Management (BLM) and the Forest Service (FS): Issues for the 111th Congress*, coordinated by Ross W. Gorte and Carol Hardy Vincent.

3 Act of March 3, 1891, ch. 561; 16 U.S.C. §471. The forest reserves were renamed national forests in the Act of March 4, 1907. The authority for the President to proclaim forest reserves was repealed in 1976; now only Congress can create new national forests or modify the existing boundaries.

4 Seventh unnumbered paragraph under "Surveying the Public Lands" in § 1 of the Sundry Civil Appropriations Act for FY1898, Act of June 4, 1897, ch. 2; 16 U.S.C. §475.

5 Act of March 4, 1907, ch. 2907.

6 Act of February 1, 1905, ch. 288; 16 U.S.C. §472.

7 P.L. 86-517; 16 U.S.C. §§ 528-531.

8 P.L. 94-588; 16 U.S.C. §§ 1600-1614. NFMA was largely an amendment to the Forest and Rangeland Renewable Resources Planning Act of 1974 (RPA; P.L. 93-378). RPA requires a periodic national assessment of resource conditions, a program for all USFS activities to address needs identified in the assessment, a presidential statement of policy, and an annual report, as well as requiring management plans for NFS units. Despite this unit planning requirement in RPA, such planning is commonly referred to as NFMA planning (or forest planning) because NFMA added many details on considerations and requirements in the planning process.

[9] The USFS is allowed to combine national forests administratively for planning purposes. Thus, while there are 155 proclaimed national forests, there are only 105 national forest units for planning and administration.

[10] P.L. 91-190; 42 U.S.C. §4321-4347.

[11] P.L. 93-205; 16 U.S.C. §§1531-1540.

[12] Citizens for Better Forestry v. U.S. Dept. of Agriculture, 481 F.Supp. 2d 1059 (N.D. Cal. 2007).

[13] Citizens for Better Forestry v. U.S. Dept. of Agriculture, 632 F.Supp. 2d 968 (N.D. Cal. 2009).

[14] For more information, and the current status of planning regulations and litigation, see CRS Report R40237, *Federal Lands Managed by the Bureau of Land Management (BLM) and the Forest Service (FS): Issues for the 111th Congress*, coordinated by Ross W. Gorte and Carol Hardy Vincent.

[15] Congress also created a National Park System and a National Wildlife Refuge System, both administered within the Department of the Interior, for special management emphasis on the federal lands included in those systems. While there are hearings and committee reports on most statutes designating special management areas, there has been very little discussion of why Congress has chosen to designate some areas for special management emphasis by the USFS and other areas for relatively similar management emphasis by a different federal agency.

[16] A third "system" with a specific purpose, the National Trails System, also contains NFS lands. However, the National Trails System Act (P.L. 90-543; 16 U.S.C. §§1241-1249) includes no direction for managing corridors along the designated routes, and thus creates no special management areas.

[17] For background, see CRS Report RL31447, *Wilderness: Overview and Statistics*, by Ross W. Gorte.

[18] P.L. 88-577; 16 U.S.C. §§ 1130-1136.

[19] Although protection of the forest reserves was an original purpose of the NFS, some early USFS employees (notably Aldo Leopold and Bob Marshall) recognized the need to protect some areas in their natural state. Acting at its own discretion, the USFS created the first wilderness area in the Gila National Forest (NM) in 1924. In the succeeding decades, the agency's system of wilderness, wild, and primitive areas (most of which have been subsequently designated as wilderness by Congress) grew to 14.6 million acres.

[20] For more detailed information on management directions and exceptions, see CRS Report RL33827, *Wilderness Laws: Permitted and Prohibited Uses*, by Ross W. Gorte.

[21] The language only directly applied to the areas designated in the 1964 act, but it has generally been applied to other designated areas by reference to the 1964 act in the subsequent laws designating wilderness areas.

[22] Many bills also contain provisions withdrawing the lands from entry appropriation, and disposal under the public land, mining, and mineral and geothermal leasing laws, effectively reinforcing Wilderness Act provisions. "No buffer zone" language is also often included to prevent the wilderness designation from affecting management of adjoining lands, even though the Wilderness Act contains no authority to regulate activities outside the designated areas.

[23] See CRS Report RL30755, *Forest Fire/Wildfire Protection*, by Ross W. Gorte.

[24] P.L. 90-542; 16 U.S.C. §§ 1271-1287.

[25] The congressionally designated game refuges and wildlife preserves within the NFS remain part of the NFS, and have not been transferred to and are not part of the National Wildlife Refuge System administered by the Fish and Wildlife Service in the Department of the Interior.

[26] P.L. 98-579; 43 U.S.C. §§ 1701 et seq. The repeal of most of the public land laws was done in several sections in Title VII of FLPMA, and generally allowed their continued application in Alaska for an additional 10 years (i.e., through 1986).

[27] Uneven-aged timber management is a silvicultural approach where a relatively few trees are harvested at relatively frequent intervals to yield timber stands with a variety of sizes (and of species, where that is natural and feasible). This contrasts with even-aged timber management, where all or the majority of trees on a site are harvested at the same time (e.g., by clearcutting) and the regenerated stand is generally trees of the same size and species.

[28] The length is provided for the statute, and sometimes for relatively long provisions in the statute, as an indication of the level of detail enacted by Congress.

[29] The Santa Rosa and San Jacinto Mountains National Monument is jointly managed with the BLM; more than 2 million acres of the 2.7 million acre monument are BLM land.

[30] Misty Fjords and Admiralty Island National Monuments were initially created by presidential proclamation, but the designations were subsequently enacted by Congress.

[31] The Transferal Area, Corridor, and Area Adjacent are unique designations on the referenced map that are to become part of the Monument upon the termination, cancellation, or relinquishment of the three specified geothermal leases.

[32] Now the Federal Energy Regulatory Commission (FERC).

[33] Uneven-aged timber management is one of the standard silvicultural approaches to timber management (the other being even-aged timber management). See David M Smith et al., *The Practice of Silviculture: Applied Forest Ecology*, 9th ed. (New York: John Wiley & Sons, 1997).

[34] This excludes six comparable areas established by presidential proclamation under the authority of the Act of August 11, 1916 (16 U.S.C. §683 et seq.): Pisgah National Game Refuge (10/17/1916); Cherokee Game Refuge (8/5/1924); Big Levels Game Refuge (7/6/0935); Noontootly National Game Refuge (7/6/1938), National Catahoula Wildlife Management Preserve (9/27/1941), and National Red Dirt Wildlife Management Preserve (9/27/1941).

[35] USDA Forest Service, *Land Areas Report (LAR)—as of Sept 30, 2009*, Washington, Table 17, http://www.fs.fed.us/land/staff/lar/2009/LAR_Table_17.html.

[36] Identified as DuNoir Basin Special Management Area in legislation subsequently introduced (but not enacted) to modify the designation.

[37] This is essentially the same provision as in §13 of NFMA (16 U.S.C. §1611).

In: National Forest System: Special Management...ISBN: 978-1-62100-513-1
Editor: Michael J. Gordon

Chapter 2

National Forest System (NFS) Roadless Area Initiatives*

Kristina Alexander and Ross W. Gorte

Summary

Roadless areas in the National Forest System (NFS) have received special attention for decades. Many want to protect their relatively pristine condition; others want to use the areas in more developed ways.

Two different roadless area policies have been offered in the last decade. On January 12, 2001, the Clinton Administration's roadless area policy established a nationwide approach to managing roadless areas in the National Forest System to protect their pristine conditions. The Nationwide Rule, as it will be called in this report, generally prohibited road construction and reconstruction and timber harvesting in 58.5 million acres of inventoried roadless areas, with significant exceptions. The Bush Administration initially postponed the effective date of the Nationwide Rule, then issued its own rule that allowed states to plan how roadless areas were managed. It issued a new rule on May 13, 2005. The State Petition Rule allowed governors to petition the Secretary of Agriculture for a special rule for managing the inventoried roadless areas in their states.

Both rules were heavily litigated, with the Ninth Circuit ultimately holding that the Nationwide Rule should be in place, and the Tenth

---

* This is an edited, reformatted and augmented version of a Congressional Research Service publication, CRS Report for Congress RL30647, from www.crs.gov, dated April 18, 2011.

Circuit still considering an appeal as to whether that rule is valid. The first action halted implementation of the Nationwide Rule in 2001, but was reversed by the Ninth Circuit. In 2003, the federal District Court for Wyoming enjoined implementation of the Nationwide Rule. This holding was dismissed as moot by the Tenth Circuit in light of the 2005 State Petition Rule. In 2006, a federal court in California enjoined the State Petition Rule, and held that the Nationwide Rule applied until the Forest Service complied with the National Environmental Policy Act and the Endangered Species Act. Instead, the Forest Service allowed governors to petition for a roadless area management rule for their states under the Administrative Procedure Act. Idaho and Colorado filed petitions under this process. The Idaho petition was approved in October 2008, and upheld by a federal court in 2011. Colorado filed and then twice amended its petition. A proposed rule for Colorado roadless areas was issued on April 15, 2011, with the Forest Service proposing more than twice as many acres to be protected at the highest level—from 257,000 acres in the state's petition to 562,200 acres in the proposed rule.

In the meantime, a new lawsuit in Wyoming led to a second injunction of the Nationwide Rule by that district court. In August 2009, the Ninth Circuit upheld the California district court ruling that the State Petition Rule was invalid and that the Nationwide Rule should be in place nationwide. The appeal of the Wyoming injunction, pending in the Tenth Circuit, could lead to a conflict between the circuits, potentially creating an issue for the U.S. Supreme Court or Congress, unless the Forest Service initiates a new rule.

Following the Ninth Circuit decision, the Secretary of Agriculture reserved decisions related to timber harvesting and road construction in roadless areas to the Secretary. Much of that authority was subsequently delegated to the Chief of the Forest Service.

The Tongass National Forest has taken its own route. At first it was included in the Nationwide Rule, but then was temporarily exempted from the rule. That temporary exemption seemed moot when the State Petition Rule came into effect, but after the Ninth Circuit ruling that the Nationwide Rule applied, it was unclear whether the Tongass was still exempt. In March 2011, a federal court ruled it was not, and that the roadless areas in the Tongass National Forest were covered under the Nationwide Rule.

INTRODUCTION

Roadless areas within the National Forest System (NFS)[1] have long received special management. Beginning in 1924, the Forest Service began protecting areas administratively as wilderness, wild, and primitive areas. In

1964, Congress enacted the Wilderness Act, creating the National Wilderness Preservation System with the administratively designated Forest Service wilderness areas and directing the Forest Service (and the National Park Service and Fish and Wildlife Service) to review the wilderness suitability of certain of their lands. Since 1964, the agencies have made numerous wilderness recommendations, and Congress acted on many of them, but some recommendations remain pending. President Clinton proposed a new rule to prohibit most road construction and timber harvesting in the remaining roadless areas of Forest Service lands.[2] Implementation was delayed, then enjoined; the injunction was overturned and another injunction was set in place. President Bush then offered a new roadless area rule (referred to in this report as the State Petition Rule), but on September 20, 2006, a court set aside the State Petition Rule and reinstated the Nationwide Rule until the Bush Administration had met certain legal requirements. The Forest Service announced it was accepting petitions from states to choose their roadless areas under the Administrative Procedure Act (APA). In August 2008 a different federal court ruled that the Nationwide Rule was contrary to law.

In August 2009, the Ninth Circuit affirmed that the State Petition Rule was passed contrary to law and that the Nationwide Rule should be in place. The Tenth Circuit, however, has pending before it a case of whether the Nationwide Rule was legally adopted. There is potential for a conflict between the two circuits as to the appropriateness of the Nationwide Rule, leaving an issue that could be resolved by the U.S. Supreme Court, congressional action, or a new rulemaking by the Forest Service. Legislation was introduced in both the House and the Senate in October 2009 (S. 1738/H.R. 3692) to establish a nationwide policy of protecting inventoried roadless areas.[3]

The Secretary of Agriculture reserved to his office decisions regarding timber harvesting and road construction in roadless areas.[4] Much of that authority was delegated to the Chief of the Forest Service, with mining decisions delegated to the Under Secretary for Natural Resources and Environment. Timber harvesting approval by the Chief is generally prohibited in roadless areas unless there is an emergency, a special use authorization, or an overall benefit to the environment.

The management of the roadless areas of the NFS is of great interest to wilderness proponents and to those who favor development of natural resources in national forests. Proponents of additional protection point to the many purposes and values the roadless areas serve, including water quality protection, backcountry recreation, and habitat for wildlife. Opponents assert that the formal congressional wilderness review and designation process sets aside adequate areas for preservation and the remaining areas should be available for timber harvesting, mining, developed recreation, and other uses.

This report focuses on the roadless areas initiatives of the Clinton and Bush administrations. It discusses the regulatory and statutory background, summarizes and provides citations for the various rules and subsequent actions, and analyzes some of the legal and policy issues in connection with the roadless areas, including the litigation related to the issue.

REGULATORY AND STATUTORY BACKGROUND

Roadless Areas: Statutory Background

The principal forest management statutes relevant to analysis of the roadless rules are the "Organic Act of 1897,"[5] the Multiple Use Sustained Yield Act of 1960,[6] and the National Forest Management Act of 1976.[7] The 1897 act directs that the national forests be managed to improve and protect the forests or "for the purpose of securing favorable conditions of water flows, and to furnish a continuous supply of timber for the use and necessities of citizens of the United States."[8] The 1897 act also authorizes the Secretary of Agriculture (Secretary) to issue regulations to "regulate their [the forest reservations'] occupancy and use and to preserve the forests thereon from destruction."[9]

Over the years uses of the national forests in addition to timber and watershed management have been allowed administratively. What constitutes the most desirable combination of uses for a forest has been hotly debated for decades. Statutorily, the Multiple Use Sustained Yield Act of 1960 (MUSYA) expressly recognizes and authorizes the *multiple use* of the forests, defined as the management of all the various renewable surface resources of the national forests "in the combination that will best meet the needs of the American people" and recognizes that "some land will be used for less than all of the resources ... without impairment of the productivity of the land, with consideration being given to the relative values of the various resources, and not necessarily the combination of uses that will give the greatest dollar return or the greatest unit output."[10] MUSYA states that the national forests are established and shall be administered for their original purposes and also "for outdoor recreation, range, timber, watershed, and wildlife and fish purposes"[11] and that "the establishment and maintenance of areas of wilderness are consistent with the purposes and provisions of" this act.[12] This latter language, which preceded enactment of the 1964 Wilderness Act,[13] recognized that the Forest Service managed some forest areas as administrative wilderness or natural areas.

MUSYA also requires *sustained yield*, defined as the "achievement and maintenance in perpetuity of a high-level annual or regular periodic output of the various renewable resources of the national forests without impairment of the productivity of the land."[14] How much is a "high-level annual or regular periodic output" of forest resources that does not impair the productivity of the land has also been the subject of much debate.

The National Forest Management Act of 1976 (NFMA) sets out additional provisions on the management of the national forests that include direction for developing, amending, and periodically revising land and resource management plans. NFMA directs that regulations be adopted to guide forest planning and accomplish specific goals set by the Congress "under the principles of" MUSYA. This includes insuring "consideration of the economic and environmental aspects of various systems of renewable resource management, including the related systems of silviculture and protection of forest resources, to provide for outdoor recreation (including wilderness), range, timber, watershed, wildlife, and fish; ... [and] provide for diversity of plant and animal communities."[15]

THE CLINTON ADMINISTRATION ROADLESS AREA RULE

The Nationwide Rule

The Clinton Administration undertook a series of actions affecting the NFS roadless areas. On October 13, 1999, President Clinton directed the Secretary of Agriculture to develop regulations to provide "appropriate long-term protection for most or all of the currently inventoried 'roadless' areas, and to determine whether such protection is warranted for any smaller roadless areas not yet inventoried."[16] In response, the Forest Service prepared an environmental impact statement (EIS) on alternatives to protect NFS roadless areas. The final rule, issued on January 12, 2001, was to take effect on March 13, 2001.[17] The Nationwide Rule established a nationwide policy for roadless areas. National-level guidance was deemed advisable because of roadless areas' importance for various forest management purposes and to the American public, and because addressing projects in roadless areas on a forest-by-forest basis as part of the planning process had resulted in a great deal of time and money spent on appeals and litigation.[18]

The Nationwide Rule (1) prohibited, with significant exceptions, new roads in inventoried roadless areas; (2) prohibited most timber harvests in the roadless areas, but allowed cutting under specified circumstances; and (3) applied those same prohibitions to the Tongass National Forest in Alaska but allowed certain road and harvest activities already under way to be completed.

The Forest Service identified approximately 58.5 million acres of inventoried roadless areas, 30% of all NFS lands. Roads are also currently prohibited in an additional 34.9 million acres of congressionally designated wilderness areas (18% of NFS lands), and are restricted in another 9.2 million acres (5% of NFS lands) of other congressionally designated areas, such as national recreation areas and wild and scenic river corridors. The roughly 47% of NFS lands with roads contain about 300,000 miles of Forest Service and other roads.[19]

The notice of final rulemaking states that roadless areas provide significant opportunities for dispersed recreation, are sources of public drinking water, and are large undisturbed landscapes that provide open space and natural settings, serve as a barrier against invasive plant and animal species, are important habitat, support the diversity of native species, and provide opportunities for monitoring and research.[20] In contrast, that notice continues, installing roads can increase erosion and sediment yields, disrupt normal water flow processes, increase the likelihood of landslides and slope failure, fragment ecosystems, introduce non-native species, compromise habitat, and increase air pollution.[21]

The final roadless area rule was more restrictive in several respects than either the proposed rule or the preferred alternative in the final EIS. With some exceptions the rule imposed immediate, national-level, Service-wide limitations on new road construction and reconstruction in the inventoried roadless areas throughout the NFS, and imposed nationwide prohibitions on timber harvesting in those areas. The regulations were to apply immediately to the Tongass National Forest in Alaska, although certain activities already in the planning stages in that forest were allowed to go forward.

The final rule prohibited new road construction and reconstruction, with exceptions for health, safety, resource protection, or other issues related to the public interest.[22] Any classified roads (those in the NFS transportation system or otherwise authorized by the Forest Service) could be maintained in inventoried roadless areas.

Harvesting timber was expected to be infrequent under the rule. Cutting, selling, or removing timber from inventoried roadless areas was prohibited unless one of specified circumstances existed. Cutting small-diameter trees was permissible to maintain or improve one or more of the roadless area characteristics and "to improve threatened, endangered, or sensitive species habitat; or to maintain or restore the characteristics of ecosystem composition and structure, such as to reduce the risk of uncharacteristic wildfire effects."[23]

Other cutting could be permitted if incidental to implementing a management activity that was not otherwise prohibited; if needed and appropriate for personal or administrative use in accordance with 36 C.F.R. § 223 (the regulations on sale and disposal of timber); or if roadless characteristics had been substantially altered in a portion of an inventoried roadless area due to the construction of a classified road and subsequent timber harvest before January 12, 2001. In this last instance, timber could only be cut in the substantially altered portion of the roadless area.[24]

The Nationwide Rule expressly did not affect any permit, contract, or other legal instrument authorizing the occupancy and use of NFS lands issued before January 12, 2001; nor did it alter any project or activity decision made prior to that date.[25] The rule did not apply to roads or harvest in the Tongass National Forest if a notice of availability of a draft EIS for the activities had been published in the *Federal Register* before January 12, 2001.[26] Otherwise the new rule applied to the Tongass National Forest immediately.

Interim Management before the Nationwide Rule Took Effect

The Nationwide Rule did not take effect immediately, except for the Tongass National Forest. Interim guidance in the Forest Service Manual (Manual) on the management of roadless areas and the construction of roads in roadless areas would have applied until a roads analysis was completed and incorporated into the relevant forest plans.[27] The Manual contained considerable detail that would have permitted new roads only if the regional forester determined there was a compelling need for the road, and both an EIS and a science-based roads analysis had been completed. The management direction was to apply both to inventoried roadless areas and to areas of more than 1,000 acres that were contiguous to inventoried roadless areas (or certain other areas) and met stated criteria. Exceptions were provided to the applicability of the interim guidelines.

Administrative Delay in Implementation

Immediately after President Bush took office, his chief of staff, Andrew Card, issued a memorandum that postponed for 60 days the effective date of regulations that were not yet in effect.[28] The roadless area regulation was covered by this language.

On February 5, 2001, notice was published in the *Federal Register* postponing the effective date of the Nationwide Rule from March 13, 2001, to May 12, 2001.[29] The Administration would then decide whether or not to implement the rule. The Nationwide Rule was enjoined by the District Court of Idaho on May 10, 2001 (see below, "Roadless Rule Litigation").

Proposed Rulemaking for a New Roadless Policy

On July 10, 2001, the Forest Service published an Advance Notice of Proposed Rulemaking, asking for public comment on 10 questions relating to "key principles" involving management of the roadless areas. Comments were due in September. The questions asked about the role of local forest planning, collaboration, wildfire prevention, access to roadless areas, and appropriate activities.

On June 26, 2002, the Forest Service released its summary report (dated May 31, 2002) on the public comments received in response to the Advance Notice of Proposed Rulemaking. The Forest Service received about 726,000 responses, mostly form letters, but also 52,432 original responses. The report urged caution in relying on the gist of the comments received, in that "respondents are self-selected; therefore their comments do not necessarily represent the sentiments of the entire population. The analysis attempts to provide fair representation of the wide range of views submitted, but makes no attempt to treat input as if it were a vote." Appendix E of the report indicated that the overwhelming number of "organized" responses were in favor of the roadless rule.[30]

Nationwide Rule Enjoined but then Reinstated

In 2001 the federal District Court for Idaho stopped implementation of the Nationwide Roadless Rule under a preliminary injunction.[31] The court found it likely that the Forest Service had not complied with the National

Environmental Policy Act (NEPA) when preparing the rule, pointing to the alternatives analysis and public comments sections of the environmental impact statement (EIS) as probably flawed.[32] At this point, the Bush Administration had stopped defending the rule, planning to withdraw it. The defense was mounted by environmental groups that intervened as defendants. They brought an appeal. The Ninth Circuit reversed the Idaho District Court.[33] The Circuit Court held that the Forest Service had complied with NEPA by analyzing a reasonable range of alternatives and providing "the public with extensive, relevant information on the Roadless Rule."[34] The Nationwide Rule was back in effect.

Nationwide Rule Enjoined for Violating the Wilderness Act and NEPA

On July 14, 2003, the federal District Court for Wyoming permanently enjoined the Nationwide Rule for violating NEPA by failing to provide adequate public notice and on the grounds that it was a "thinly veiled attempt to designate 'wilderness areas' in violation of the clear and unambiguous process established by the Wilderness Act."[35] The court equated the roadless areas with de facto wilderness, characterizing the severity of the restrictions under the roadless rule as restrictive as for congressionally designated wilderness areas.[36] The court concluded that the rule was "in violation of the clear and unambiguous process established by the Wilderness Act for such Designation."[37]

Only Congress can designate areas for inclusion in the National Wilderness Preservation System.[38] However, MUSYA—enacted before the 1964 Wilderness Act—expressly provides for the administrative management of national forest lands for many purposes, and states that "the establishment and maintenance of wilderness areas is consistent with" the purposes of the act. NFMA—enacted after the Wilderness Act—directs that forest plans "assure ... coordination of outdoor recreation, range, timber, watershed, wildlife and fish, and *wilderness*" (emphasis added). Therefore, it appears that as a general matter, the roadless rule might be defended as appropriate management of non-timber resources for multiple use purposes (such as outdoor recreation, water quality protection, mineral development, and game and other wildlife habitat). However, it also is possible that severe and extensive restrictions might be seen as violating sustained yield under MUSYA.

The Nationwide Rule included explanatory material that distinguished roadless areas from wilderness areas, stating that other management is still allowed—such as development of mineral claims, livestock grazing, off-highway vehicle use, and fire management efforts.[39]

The Wyoming court did not address the provisions for using NFS as wilderness under either the MUSYA or NFMA. The validity of the court's assertion, that the roadless rule was "in violation of the clear and unambiguous process established by the Wilderness Act for such [Wilderness] Designation,"[40] would seem to depend on whether one agrees that management of the roadless areas under the roadless rule would be as restrictive as that under the Wilderness Act, and on how one views the references to wilderness in MUSYA and NFMA and the actions of Congress in continuing to allow management of the roadless areas to maintain wilderness characteristics on national forest lands.

Management of Roadless Areas after Nationwide Rule Was Enjoined

After the Nationwide Rule was enjoined, the Bush Administration issued a series of interim directives governing roadless area protection and management until it could formulate its own rule. However, the interim directives were only to be in effect for 18 months and would apparently cease to apply once a forest plan was revised or amended.

A December 2001 directive, ID 1920-2001-1, appeared to replace many of the previous interim directives (both Clinton and Bush). However, the *Federal Register* notice did not clearly indicate which provisions were being replaced or the precise extent of the revisions. The published explanatory material stated that affected material was shown and unaffected material was not. Yet some of the earlier provisions were neither shown nor discussed and, therefore, might still have been in effect. The final text of new Manual § 1925 did not show these undiscussed earlier provisions—as though they were superseded. Thus, it was not clear which of the previous materials were still in effect. For example, some of former Manual § 7712.16 (that contained specific details on permissible road construction) was expressly revised in the December directives, and the explanatory materials stated that the revised provisions were moved to the Manual as a new § 1925. Yet other provisions that were in § 7712.16 were neither discussed as superseded or modified, nor set out in the new § 1925. Thus, it was not clear exactly which of the previous interim directives were still in effect after the December directive took effect.

Decision-Making on Timber Harvesting in Roadless Areas

The December directive reserved authority to the Chief of the Forest Service to approve certain proposed timber harvests in inventoried roadless areas until a forest plan was completed "that has considered the protection and management of inventoried roadless areas pursuant to FSM 1920." They also provided that the Chief could designate an Associate Chief, Deputy Chief, or Associate Deputy Chief, on a case-by-case basis, to be the responsible official.

Regional foresters were to screen timber harvest projects in inventoried roadless areas for possible referral to the Chief. The Chief was to make decisions regarding harvests *except* for those that were (1) generally of small-diameter material, the removal of which is needed for habitat or ecosystem reasons (including reducing fire risk); (2) incidental to a management activity not prohibited under the plan; (3) needed for personal or administrative use; or (4) in a portion of an inventoried roadless area where harvests had previously taken place and the roadless characteristics had been substantially altered. (These exceptions conform to the timber harvest exceptions in the Nationwide Rule.) Decisions as to these harvests were to be made by forest officers normally delegated such authority.

The December 2001 directive stated that the Chief's authority with respect to timber harvests did not apply if a Record of Decision for a forest plan revision was issued as of July 27, 2001—as was true of the Tongass National Forest—and would otherwise terminate when a plan revision or amendment that has considered the protection and management of inventoried roadless areas was completed.

Decision-Making on Road Construction in Roadless Areas

The Chief's authority for road construction in inventoried roadless areas was to remain in effect until a forest-scale roads analysis was completed and incorporated into each forest plan.[41] Regional foresters were to make many decisions on road construction projects under the new Manual § 1925.04b. There was no express provision for terminating the authority of regional foresters, but the general policy, § 1925.03, keyed termination of the special provisions to completion of a roads analysis and its incorporation into the relevant forest plan.

The December Directive apparently eliminated the Clinton-era requirement that there be a *compelling* need for a road, a science-based analysis, and full EIS in all cases. The applicability of the interim directive to certain contiguous areas also was eliminated. The responsible official could still do an EIS and could protect contiguous areas, and could find a compelling need for a road. However, roadless areas could have reduced protections because the directives did not contain the higher thresholds for approval of activities and more formalized documentation requirements previously required.

On July 16, 2004, the Forest Service reinstated the December 2001 directive ID 1920-2001-1 (which had expired on June 14, 2003) simultaneously with a proposed new roadless rule.[42] It was renumbered as ID 1920-2004-1, and modified. It reestablished the provisions allowing the Chief to make decisions affecting inventoried roadless areas involving (1) road construction, until a forest-scale roads analysis was completed and incorporated into a forest plan or a determination was made that a plan amendment was not needed; and (2) timber harvests, until a forest plan had "considered" protecting and managing roadless areas.

ID 1920-2004-1 differs from the 2001 version by allowing the Chief to grant project-specific exceptions to allow a regional forester or forest supervisor to exercise the authority to conduct projects in roadless areas. Adding forest supervisors to those who can decide to conduct projects in roadless areas expanded the 2001 directive, and arguably made it easier to approve projects in those areas. Manual § 1925.04b also was changed to allow regional foresters to make decisions on road construction and reconstruction in inventoried roadless areas for lands associated with any mineral lease, license, permit, or approval for mineral leasing operations.

Dispute over Tongass National Forest Roadless Areas

The Nationwide Rule was scheduled to apply immediately to the Tongass National Forest in Alaska. The state of Alaska sued, arguing that the Nationwide Rule abrogated the Alaska National Interest Lands Conservation Act of 1980 (ANILCA), which provided there would be no additional administrative withdrawals of federal lands in the state. To settle the lawsuit, the Bush Administration agreed it would publish a proposed roadless rule that would exclude the Tongass National Forest.[43] A proposed rule to exclude the Tongass from the Nationwide Rule and an Advance Notice of Proposed Rulemaking to exclude both the Tongass and Chugach National Forests (also in Alaska) were published on July 15, 2003.[44] A final rule "temporarily" exempting the Tongass from the roadless rule was published on December 30, 2003.[45] Specifically, the roadless area restrictions on road construction and reconstruction and on the cutting, sale, or removal of timber in inventoried roadless areas did not apply to the Tongass National Forest. The explanatory material indicated that the exemption was in place only until an Alaska-wide roadless rule was finalized. The exemption meant that areas protected under the Revised Tongass Land Management Plan would remain protected, but an additional 300,000 acres of inventoried roadless areas would be available for logging under that plan.

The Forest Service has taken the position that the State Petition Rule, published on May 13, 2005 (discussed in "2005 Roadless Area Policy: The State Petition Rule," below), eliminated the need for further Tongass-specific rulemaking and that timber harvest decisions that include an inventoried roadless area component will be made in accordance with the forest plan unless changed through state-specific rulemaking.[46] A new Tongass Forest plan amendment that would make some roadless areas available for logging was completed in January 2008; the Record of Decision was appealed administratively, but the plan was upheld by Chief Gale Kimball on August 22, 2008. A court upheld the 2008 Tongass Land and Resource Management Plan in 2010.[47]

However, in 2011 a federal court ruled that the Nationwide Rule applied to the Tongass National Forest. The court held that the 2003 decision putting the "temporary" Tongass exemption into place violated the Administrative Procedure Act (APA).[48] The court found that the justifications for exempting the Tongass from the roadless rule, such as maintaining jobs, allowing development of community roads and utility corridors, or that existing plans protected the roadless areas, were contradicted or otherwise not supported by other findings of the Forest Service.

2005 Roadless Area Policy: The State Petition Rule

New final roadless area rules were published on May 13, 2005, giving power to the states to request which roadless areas would remain unchanged.[49] The changes to 36 C.F.R. § 394 permitted the governor of a state to recommend management of roadless areas in the state in a petition to the Secretary. Petitions were required to provide specified information, including

(1) a description of the particular lands for which the petition is made; (2) particular management recommendations; (3) identification of the needs and circumstances intended to be addressed by the petition, including conserving roadless value, protecting health and safety, reducing fuels, maintaining dams, and providing access; (4) information on how the recommendations differ from the relevant federal plans and policies; (5) a comparison of the recommendations with state land management policies and direction; (6) impact of the recommendations on fish and wildlife and habitat; (7) a description of any public involvement efforts undertaken by the state in developing the petition; and (8) a commitment by the state to participate as a cooperating agency in any environmental analysis for the rule-making.

Petitions were allowed until November 13, 2006, with provisions for late petitions under 7 C.F.R. § 1.28. If the Secretary approved a petition, the Forest Service must coordinate development of the proposed rule with the state, but the Secretary would make the final decision on any rule. The rationale for imposing a deadline on the roadless area petition process, and the consequences of a state's failure to meet the deadline, are unclear. The rule does not indicate what happens when there is a change in governors. This ambiguity became focused when the new governor of Colorado disagreed with the designations made in the petition submitted by his predecessor shortly after the prior governor lost re-election. See "State Petition Submissions and Responses," below.

The rule does not include any standards on the scope or balance required for public participation in developing a state's recommendations, or for how a state is to gather its information about an area's resources and values. There also are no requirements or standards for the Secretary's review and approval of a petition, and no statement as to the relative weight to be given to a state's recommendations versus the preferences of non-state residents who may express an opinion on the desired management of these national lands. An advisory committee was established to provide advice on implementing the rule.[50]

The background materials accompanying the proposed and final rule refer to the importance of collaborating with partners, including state governments: "strong State and Federal cooperation regarding management of inventoried roadless areas can facilitate long-term, community-oriented solutions."[51] The materials review the promulgation of the Nationwide Rule and note that the new rule was proposed in response to "concerns [that] were immediately expressed by those most impacted by the roadless rule's prohibitions."[52] This latter comment raises questions about the role of public input in the rulemaking in that only 4% of the comments received on the original roadless rule opposed or expressed concerns about its protections. Similarly, the responses to the new roadless rule overwhelmingly favored continuing protection. The explanatory materials justify taking a different approach by stating that "the public comment process is not intended to serve as a scientifically valid survey process to determine public opinion."[53]

State Responses to the State Petition Rule

Response to the State Petition Rule by governors was mixed. Most states that filed petitions sought to preserve all of their roadless areas, providing the same protection as under the Nationwide Rule.[54] On January 12, 2004, nine governors wrote in opposition to the rule.[55] Other governors expressed concern that the new roadless area rule petition process was too vague to provide states proper guidance in proposing state-based rules, and placed undue burdens on state resources. North Carolina Governor Michael Easley raised the vagueness issue in a letter to the Secretary of Agriculture, asserting that the rule provided "insufficient clarity" on how the department would respond to state petitions or the criteria the department would use to consider them.[56] He also claimed that the state administrative and financial investment in the petitioning process could be "onerous."

Other governors also expressed concern over the rule's burden on states. On October 14, 2004, Oregon Governor Theodore Kulongoski filed a petition with Secretary Johanns to permit states to adopt the 2001 roadless area rule through an expedited process. The governor argued that the 2005 roadless area rule would require states to expend significant financial and personnel resources to ensure adequate public participation and technical analysis on its provisions and implementation.[57] Tennessee Governor Bredesen similarly stated in a September 8, 2004, letter to Secretary Veneman that the rule may place an "undue financial burden on our state agencies without providing some assurances of any concurrence or approval from the Department of Agriculture."[58] Virginia Governor Mark Warner expressed concern in a July 30, 2004, letter to the Secretary that, although the 2005 rule confirmed that the Forest Service had final authority for managing roadless areas, the rule placed the burden of seeking protection of the areas on individual states.[59]

State Petition Submissions and Responses

Governors of five states filed petitions with the Secretary of Agriculture under the State Petition Rule. Beginning with Virginia, on December 22, 2005, the governors of North Carolina, South Carolina, New Mexico, and California submitted petitions requesting comprehensive protection of all roadless areas in their respective states. The governor of Idaho filed a petition for managing Idaho's inventoried roadless areas on October 5, 2006, which plan would include resource development in some areas.[60] The lame-duck governor of Colorado filed a petition on November 13, 2006, to keep a significant portion of the areas roadless. An amended petition was filed by his successor in April 2007, recommending preservation of even more acreage.[61] Both the Idaho and the Colorado petitions were made after the State Petition Rule was enjoined (see "Implementation of the State Petition Rule Enjoined," below).

The advisory committee recommended that the petitions of North Carolina, South Carolina, and Virginia be approved, and on June 21, 2006, the Secretary of Agriculture approved them.[62] The advisory committee did not make recommendations on the petitions of New Mexico and California, as the deadlines for responding to these petitions were interrupted by the injunction halting application of the State Petition Rule (see below).

In response to the September 20, 2006, injunction (see below), the Bush Administration announced it would process state petitions to manage roadless areas under the APA. On October 4, 2006, the charter for the advisory committee was formally amended to direct the committee's review of state petitions under the APA (5 U.S.C. § 553(e)) and under 7 C.F.R. § 1.28.[63] Both Idaho and Colorado submitted petitions under this procedure. The committee recommended approval of the Idaho and Colorado petitions. The Idaho roadless rule became final October 16, 2008.[64] The rule was challenged as violating the Endangered Species Act (ESA) and NEPA, alleging that the impacts on endangered species would be greater than described. However, the court held that the Forest Service had complied with those acts in approving the Idaho plan.[65] A proposed rule was issued for the Colorado areas in July 2008,[66] but Colorado submitted a modified petition in April 2010.[67] The Forest Service has not issued a final rule on this petition, although a proposed rule was announced in April 2011 significantly increasing the acreage designated for the highest protection from the state's petition (from 257,000 acres to 562,200 acres).[68]

ROADLESS RULE LITIGATION

Implementation of the Nationwide Rule Enjoined: Part 1

The Kootenai Tribe of Idaho, a forestry corporation, livestock concerns, motorized recreation groups, and others, sued to stop the Nationwide Rule for violating NEPA, NFMA, and the APA.[69] The state of Idaho sued the next day, seeking similar relief.[70] The federal district court deciding both Idaho cases found that plaintiffs were likely to succeed on their assertion that the Forest Service had not provided the public an opportunity to comment meaningfully on the rule because of inadequacies in (1) identification of the inventoried roadless areas (noting that statewide maps were not made available until after the public comment period had ended); (2) information presented during the scoping process (Forest Service employees were alleged to be ill-prepared); and (3) the period for public comment. The court characterized the comment period as "grossly inadequate" and an "obvious violation" of NEPA.[71] The court further found that the final EIS did not consider an adequate range of alternatives, since all but the "no action" alternative included "a total prohibition" on road construction and the EIS did not analyze whether other alternatives might have accomplished protection of the environmental integrity of the roadless areas.

The Bush Administration asked the court to postpone putting the preliminary injunction into effect until the Forest Service had reviewed the Nationwide Rule, arguing that an injunction was not necessary because the rule would not be implemented until May 12, 2001. On May 10, 2001, the court granted a preliminary injunction preventing implementation of the Nationwide Rule.[72] The court found that the government's "vague commitment" to propose amendments to the rule indicated a failure to take the requisite "hard look" that NEPA requires, leaving the court with the impression that implementation of the roadless rule would result in irreparable harm to the national forests. The court concluded that the government's response was a "band-aid approach" and enjoined implementation of the rule while the agency conducted its new study and developed amendments.

While United States did not appeal this decision, environmental groups who had intervened did.

Ninth Circuit Decision Reinstates Nationwide Rule after First Injunction

The Ninth Circuit combined the two Idaho cases on appeal, and on December 12, 2002, reversed the Idaho district court, stating: "Because of its incorrect legal conclusion on prospects of success, the district court proceeded on an incorrect legal premise, applied the wrong standard for injunction, and abused its discretion in issuing a preliminary injunction."[73]

In reaching its conclusion, the Ninth Circuit reviewed the substantive grounds considered by the district court and disagreed that plaintiffs had demonstrated a likelihood of success on the merits. It found that the Forest Service adequately complied with the NEPA public comment provision because the maps provided did not suffer from the grave inadequacies alleged by the plaintiffs. According to the Ninth Circuit, the plaintiffs had actual notice as to the roadless areas that would be affected. The court also found that the Forest Service had provided adequate time for comment,[74] and that the EIS considered an adequate range of alternatives.[75] The Ninth Circuit denied Idaho's petition for rehearing, and the case was remanded to the district court to reconsider.

Some disagreed with the Ninth Circuit decision, arguing it was improper to allow the intervenors to bring an appeal when the government chose not to. The case came forward in an unusual context. The district court decision focused on the inadequacy of the federal defendants' NEPA compliance, but the defendants did not appeal that issue. Instead, certain environmental groups that had intervened on the government's side appealed the district court's ruling. The Ninth Circuit decision raised significant issues about whether intervenors could appeal NEPAcompliance rulings when the federal defendants—the only ones who could comply with NEPA— did not.

Implementation of the Nationwide Rule Enjoined: Part 2

On July 14, 2003, the federal District Court for Wyoming permanently enjoined the Nationwide Rule because of NEPA defects and because the court concluded it created de facto wilderness in violation of the Wilderness Act[76] (see "Nationwide Rule Enjoined for Violating the Wilderness Act and NEPA," above.) In addition to the Wilderness Act conflict, the court found the Forest Service had committed the following NEPA defects in the rulemaking:

- not extending the scoping comment period;
- denial of cooperating agency status to Wyoming and other states;
- failure to rigorously explore and objectively evaluate reasonable alternatives to the Roadless Rule;
- failure to adequately analyze cumulative effects; and
- not issuing a supplemental EIS in light of new information on updated roadless area inventories.[77]

The court stated that the agency drove the rule "through the administrative process in a vehicle smelling of political prestidigitation"[78] in its "rush to give President Clinton lasting notoriety in the annals of environmentalism"[79]—and concluded that the agency must "start over."[80]

On July 11, 2005, the Tenth Circuit dismissed the appeal in this case and vacated the district court's decision. It held that the case was made moot by the new roadless regulations published by the Bush Administration on May 13, 2005.[81] Because the Tenth Circuit dismissed the litigation as moot, and the Tongass case was settled, as a principle of legal precedence, the opinion of the Ninth Circuit was the only precedent on the Nationwide Rule. Another lawsuit changed that.

Implementation of the State Petition Rule Enjoined

On August 28, 2005, the states of California, Oregon, and New Mexico filed a lawsuit in the federal District Court for Northern California challenging the State Petition Rule.[82] Environmental groups also filed suit to challenge the rule, and the cases were consolidated. The state of Washington joined the states' suit as a plaintiff intervenor, and other states filed amicus briefs on both sides.[83] The plaintiffs alleged that the State Petition Rule offered less protection and a more localized approach than the Nationwide Rule, and thus required environmental analysis under NEPA and consultation under ESA. They argued that the department had violated the APA when it finalized the regulations without conducting requisite NEPA analysis, allowing meaningful public involvement, or explaining the rationale for the department's change in roadless area policy.[84]

In a summary judgment effective September 20, 2006, the federal District Court for Northern California set aside the State Petition Rule and reinstated the Nationwide Rule until the Forest Service complied with NEPA and ESA.[85] The Forest Service appealed the decision. On November 29, 2006, the court clarified its injunction, prohibiting the Forest Service from taking any management activities that would have been banned under the Nationwide Rule.[86] This had the effect of nullifying oil and gas leases that had been issued under the State Petition Rule, among other things. The decision sparked a dispute between circuits as to which roadless rule had passed muster.

Wyoming Redux: The Nationwide Rule Enjoined: Part 3

After the California suit had been filed in 2005, the state of Wyoming attempted to revive the 2003 Wyoming District Court holding that the Nationwide Rule was invalid.[87] The Tenth Circuit issued a mandate directing the District Court of Wyoming to dismiss the 2005 challenge, which it did. Two days after the district court in California enjoined the State Petition Rule, the state of Wyoming again moved to reinstate the 2003 decision that the Nationwide Rule was contrary to NEPA and the Wilderness Act. While that 2003 case had been appealed, the Tenth Circuit had not ruled on its merits but instead found the State Petition Rule rendered it moot. Wyoming argued that the 2003 decision was still good law. The federal District Court of Wyoming refused to consider Wyoming's action to revive the holding, holding it lacked the authority.[88] However, the court stated it was "extremely perplexed" at how the California court could effectively overturn its 2003 ruling: "This court is troubled and questions the authority of the California court to raise this rule back to life and force."[89]

The state of Wyoming filed a new suit in district court, claiming the Nationwide Rule violated NEPA and the Wilderness Act. The District Court of Wyoming agreed.[90] The court considered that the Nationwide Rule would allow certain uses of the roadless areas, such as grazing, oil and gas development, and motorized vehicles, that were not traditionally allowed in wilderness. However, the court distinguished those uses, saying that to be effective, those uses all would require roads and the Nationwide Rule prevented that. The court acknowledged the California decision, saying that "the Court is disturbed, and frankly shocked at the fact that a Magistrate Judge essentially reinstituted a policy that was not properly before that Court, and especially in light of the fact that an Article III judge had already ruled that the re-instituted policy was promulgated in violation of law."[91] Because the California court had not considered the merits of the Nationwide Rule when acting (but according to the Wyoming court had instead followed a general rule that invalidating one rule reinstates its precedent), the Wyoming court said its actions were consistent with that order. The Nationwide Rule was enjoined, again.

Following the Wyoming decision, it appeared that the Forest Service was required to follow the Nationwide Rule by the California order, and forbidden to follow it by the Wyoming order. The Forest Service filed motions in both courts to suspend the injunctions so it would not be held in contempt. However, some noted that the Forest Service had not been following the Nationwide Rule as directed by the California court, but was allowing states to file petitions (as was the goal of the State Petition Rule) under the APA instead.

Judicial Comity

As a result of the Forest Service motion before the district court in California, the judge issued a rule in December 2008 limiting the scope of the injunction that reinstated the Nationwide Rule.[92] That order put the Nationwide Rule in effect in all the states within the Ninth Circuit (Alaska, Arizona, California, Hawaii, Idaho, Montana, Oregon, Nevada, and Washington) and in New Mexico. New Mexico was included because the state was a plaintiff in the case and had won the injunction. The judge stated that the notion of judicial comity, which is a concept under which courts will respect the decisions of other courts, did not *require* her to modify the injunction. She noted that her order was issued years before the Wyoming decision; the plaintiff in the Wyoming decision (the state of Wyoming) only alleged harm to the one state, but the decision was applied nationwide; and the Wyoming decision disregarded the Ninth Circuit's analysis in support of the Nationwide Rule in *Kootenai*.[93] However, she limited the injunction in the "spirit of comity."[94]

Ninth Circuit Reinstates Nationwide Rule, Again

The Ninth Circuit affirmed the lower court's decision that the State Petition Rule did not comply with NEPA or ESA and that the Nationwide Rule was in place.[95] This should have the effect of nullifying the State Petition Rule nationwide and returning the Nationwide Rule to effect at least for a while. However, the Tenth Circuit still has an appeal of the Wyoming decision pending before it. That court could exercise judicial comity and apply the reasoning of the Ninth Circuit. Even if the Tenth Circuit recognizes the sufficiency of the NEPA review of the Nationwide Rule as found by the Ninth Circuit, it could still find the Nationwide Rule invalid based on the different legal grounds presented by the wilderness argument, which the Ninth Circuit did not consider. This could lead to a conflict between the two circuits with one court saying the Nationwide Rule is valid nationwide, and another saying the contrary; this conflict between courts could be resolved only by the U.S. Supreme Court.

The conflict between rulemakings could be resolved statutorily by Congress or it could be resolved administratively by new rulemaking. The Forest Service still has a role. It could create its own rule; after all, this entire dispute is over two rulemaking procedures. A new, different roadless rule would not be constrained by these court decisions. Additionally, rather than a new rulemaking that applies for every state, the Forest Service could again use APA rulemaking on a state-by-state basis as its roadless policy, rather than follow the Nationwide Rule or the courts. In the meantime, the Forest Service instituted a one-year policy in May 2009 that only the Secretary may approve projects in roadless areas.[96] This authority was renewed in May 2010, with some timber harvesting project approval being delegated to the Chief of the Forest Service (primarily in the case of emergency or to comply with special use permits) and approval of projects regarding mining delegated to the Under Secretary for Natural Resources and the Environment.[97]

The Ninth Circuit Decision and the Tongass

Reinstatement of the Nationwide Rule by the Ninth Circuit may have reignited the conflict regarding the Tongass National Forest in Alaska: the Ninth Circuit rule did not exempt the Tongass when it declared the Nationwide Rule was properly in place, and no Tongass rule had ever been finalized, as was required by the court settlement in 2003 (see "Dispute over Tongass National Forest Roadless Areas," above). A plan was adopted that would allow development in some of the areas of the Tongass that are roadless. A court considering a challenge to the 2008 Tongass Land and Resource Management Plan did not consider the roadless rule in upholding the plan.[98]

However, the federal court for the District of Alaska held that the Forest Service violated the Administrative Procedure Act (APA) when it issued the temporary rule exempting the Tongass National Forest from the Nationwide Rule.[99] The court held it was arbitrary and capricious for the rule to find that roadless areas in the Tongass would lead to approximately 900 lost jobs, when other data of the Forest Service showed that job losses were due to market changes, not the Nationwide Rule. Other justifications for the rule were held to be arbitrary and capricious, such as that exempting the Tongass would allow community road and utility development, when the court found that such development would be allowed even under the Nationwide Rule. The court vacated the exemption but refused to nullify timber sales that were under way.

Congressional Action

On October 1, 2009, two bills were introduced to make the Nationwide Rule federal law (H.R. 3692/S. 1738 (111th)). The House bill would have directed that the final rule of January 12, 2001, be put in place and that the roadless areas shown in the final environmental impact statement for that rule be preserved. The Senate bill described the purpose of the proposed legislation, noting that road maintenance is extremely expensive and that oil and gas production on Forest Service lands accounts for 0.4% of domestic production. It defined an inventoried roadless area as any area identified as "roadless" on a map. Thus, the two bills could have affected different areas. For example, the approximately 400,000 acres of lands that Idaho decided to develop for general forest purposes pursuant to their state petition were shown as roadless when the final Nationwide Rule was published.[100] The House bill appeared to have the potential effect of nullifying the Idaho state plan, whereas the Senate bill might have affected only some of the areas in the Idaho rule.

CONCLUSION

Roadless areas of the National Forest System have received special attention for decades. Some argue that the areas should be available for appropriate development—mineral exploration and development, timber harvesting, motorized recreation, developed recreation sites, and more. Others believe that the areas should remain roadless to preserve the special values that their condition provides—clean water, undeveloped wildlife habitats, dispersed recreation, aesthetics, and more.

Both the Bush Administration and the Clinton Administration issued rules addressing inventoried roadless areas—one rule in which the states would take the lead in management decisions, and the other applying a uniform national policy. Both were rejected by courts, leading to conflicting decisions in two different district courts—one reinstating the Nationwide Rule, and the other enjoining it—and a circuit court decision reinstating the Nationwide Rule and nullifying the State Petition Rule.

The Forest Service under the Obama Administration has indicated that most projects in roadless areas must be approved by the Secretary, suggesting it is not continuing the Bush-era policy of allowing governors to file petitions for a state-specific rule for managing roadless areas under the APA, nor does it appear to be following the Nationwide Rule which prohibits projects in roadless areas. Only one state—Colorado—has a pending roadless rule application under the APA policy.

The ultimate decision of which roadless rule applies may not be in the courts, unless it is brought before the Supreme Court, but by Congress. The contradictory court decisions may indicate a statutory fix is needed. Legislation was introduced in the 111th Congress in both houses to make the Nationwide Rule federal law.

End Notes

[1] The NFS includes 155 national forests (188.0 million acres), 20 national grasslands (3.8 million acres), and 125 other units (0.9 million acres), and is administered by the Forest Service in the U.S. Department of Agriculture.

[2] The *inventoried roadless areas*, identified in the Nationwide Rule and most other documents, generally refer to areas identified in the Forest Service's second *Roadless Area Review and Evaluation* (RARE II) completed in January 1979, excluding areas that have since been designated as part of the National Wilderness Preservation System by Congress.

[3] H.R. 3692/S. 1738 (111th Cong.). For more on current legislative activity, see CRS Report R40237, *Federal Lands Managed by the Bureau of Land Management (BLM) and the Forest Service (FS): Issues in the 111th Congress*, coordinated by Ross W. Gorte and Carol Hardy Vincent.

[4] Secretary's Memorandum 1042-155, Authority to Approve Road Construction and Timber Harvesting in Certain Lands Administered by the Forest Service (May 28, 2010), available online at http://www.fs.usda.gov/Internet/FSE_DOCUMENTS/stelprdb5167211.pdf.

[5] The 3rd through 17th unnumbered paragraphs under "Surveying the Public Lands," in the Sundry Civil Expenses Appropriations Act for FY1898; Act of June 4, 1897 (ch. 2, 30 Stat. 34), 16 U.S.C. §§ 473-475.

[6] P.L. 86-517, 74 Stat. 215; 16 U.S.C. §§ 528-531.

[7] P.L. 94-588, 90 Stat. 2949, primarily amending P.L. 93-378, the Forest and Rangeland Renewable Resources Planning Act of 1973 (RPA); 16 U.S.C. §§ 1600-1614.

[8] 16 U.S.C. § 475.

[9] 16 U.S.C. § 551.

[10] 16 U.S.C. § 531.

[11] 16 U.S.C. § 528.

[12] 16 U.S.C. § 529.

[13] P.L. 88-577, 78 Stat. 890; 16 U.S.C. §§ 1131-1136.

[14] 16 U.S.C. § 531.

[15] 16 U.S.C. § 1604(g). Note that Forest Service wilderness management is again mentioned in law, 12 years after enactment of the Wilderness Act.

[16] Memorandum from President William J. Clinton to the Secretary of Agriculture, *Protection of Forest 'Roadless' Areas* (October 13, 1999).

[17] 66 Fed. Reg. 3244 (January 12, 2001), adding 36 C.F.R. § 294, Subpart B.

[18] 66 Fed. Reg. at 3246.

[19] The Forest Service does not report its inventory of roads, but identified that the 64,866 miles of road maintained to standard was 22% of all roads. U.S. Department of Agriculture, *Forest Service Performance and Accountability Report—Fiscal Year 2004* (April 2005).

[20] 66 Fed. Reg. 3245 (January 12, 2001).

[21] 66 Fed. Reg. at 3246.

[22] See 66 Fed. Reg. 3272; 36 C.F.R. § 294.12(b).

[23] 36 C.F.R. § 294.13(b)(1).

[24] 36 C.F.R. § 294.13(b)(2)-(4).

[25] 36 C.F.R. § 294.14(a) and (c).

[26] 36 C.F.R. § 294.14(d).

[27] 66 Fed. Reg. 3219.

[28] Andrew H. Card, Jr., Memorandum for the Heads and Acting Heads of Executive Departments and Agencies (January 20, 2001).

[29] 66 Fed. Reg. 8899. The postponement notice stated that the action was exempt from notice and comment because it was a procedural rule and for good cause shown.

[30] The Report on the Public Comments is available online at http://www.roadless.fs.fed.us. The Ninth Circuit pointed out that the Attorney General of Montana had asserted that nationally, "96% of commenters favored stronger protections." Kootenai Tribe of Idaho v. Veneman, 313 F.3d 1094 (9th Cir. 2003).

[31] Kootenai Tribe of Idaho v. Veneman, 2001 WL 1141275 (D. Idaho, May 10, 2001).

[32] Kootenai Tribe of Idaho v. Veneman, 142 F. Supp. 2d 1231 (D. Idaho 2001).

[33] Kootenai Tribe of Idaho v. Veneman, 313 F.3d 1094 (9th Cir. 2002).

[34] Kootenai Tribe of Idaho v. Veneman, 313 F.3d 1094, 1116 (9th Cir. 2002).

[35] Wyoming v. U.S. Department of Agriculture, 277 F. Supp. 2d 1197, 1239 (D. Wyo. 2003).

[36] The court reviewed the rule's exceptions allowing roads in roadless areas, but failed to mention that Federal Aid Highways could be permitted in some instances (p. 1236). On the other hand, the court noted that the roadless rule was more restrictive for constructing roads in roadless areas to combat problem conditions than was the Wilderness Act, which allows necessary measures to control fire, insects, and diseases, while the roadless rule only allows roads in the case of an "imminent flood, fire, or other catastrophic event that, without intervention, would cause the loss of life or property."

[37] Wyoming v. U.S. Department of Agriculture, 277 F. Supp. 2d at 1239.

[38] 16 U.S.C. § 1131.

[39] 66 Fed. Reg. 3249 (January 12, 2001).

[40] Wyoming v. U.S. Department of Agriculture, 277 F. Supp. 2d at 1239.

[41] 66 Fed. Reg. 65801.

[42] 69 Fed. Reg. 42648.

[43] Alaska v. U.S. Department of Agriculture, No. A01-039 CV (D. Alaska June 10, 2001).

[44] 68 Fed. Reg. 41865 and 68 Fed. Reg. 41864 (July 15, 2003), respectively.

[45] 68 Fed. Reg. 75136 (December 30, 2003).

[46] See http://www.fs.fed.us/r10/tongass/forest_facts/faqs/roadless.shtml.

[47] Southeast Conference v. Vilsack, 684 F. Supp. 2d 185 (D.D.C. 2010).

[48] Organized Village of Kake v. U.S. Department of Agriculture, 1:09-cv-00023 JWS (D. Alaska March 4, 2011).

[49] 70 Fed. Reg. 25654.

[50] 70 Fed. Reg. 25663.

[51] 70 Fed. Reg. 25654.

[52] Id.

[53] 70 Fed. Reg. 25656.

[54] States that filed to protect all of their roadless areas are: California, New Mexico, North Carolina, South Carolina, and Virginia. Petitions from

Colorado and Idaho recommended development in some of the inventoried roadless areas.

[55] The governors were from the following states: Arizona, Iowa, Kansas, Maine, Missouri, New Mexico, Pennsylvania, Virginia, and Washington.

[56] Letter to Ann Veneman, U.S. Secretary of Agriculture (August 16, 2004).

[57] Letter from Theodore Kulongoski, Governor of Oregon, to Mike Johanns, U.S. Secretary of Agriculture (October 14, 2005). On October 27, 2005, the department denied the petition for an expedited process scheme, expressing doubt that the Governor's proposed rule would substantially decrease states' burdens. Letter from Mark Rey, Under Secretary of Natural Resources and Environment, U.S. Department of Agriculture, to Theodore Kulongoski, Governor of Oregon.

[58] Letter to U.S. Secretary of Agriculture Ann Veneman (September 8, 2004).

[59] Letter to U.S. Secretary of Agriculture Ann Veneman (July 30, 2004). The Forest Service provided grants to states that requested financial assistance for petition development: $150,000 to Idaho (December 20, 2005); and $200,000 to the State of Arizona (June 28, 2006). U.S. Dept. of Agriculture Press Releases Nos. 0560.05 and 0225.06.

[60] Petition of Governor James E. Risch for Roadless Area Management in Idaho, available online at http://www.fs.fed.us/emc/roadless/061005_gov_risch_petition.pdf.

[61] The Roadless Area National Advisory Committee recommended approving the April 2007 petition. Daily Env't Rep. (BNA), p. A-7 (August 13, 2007). On August 3, 2009, Colorado revised its roadless application. It was revised again in April 2010. See http://www.dnr.state

[62] USDA News Release No. 0212.06.

[63] 71 Fed. Reg. 58577 (October 4, 2006).

[64] 73 Fed. Reg. 61455 (October 16, 2008). 36 C.F.R. part 294. The Idaho final rule created five different management areas within its 9.3 million acres that would prohibit some forms of development, depending on the classification, except in the 405,900 acres designated as general forest.

[65] Jayne v. Rey, No. 4:09-CV-015-BLW (D. Idaho January 29, 2011).

[66] 73 Fed. Reg. 43543 (July 25, 2008).

[67] See Colorado Department of Natural Resources website at
http://www.dnr.state

[68] 76 Fed. Reg. 21272 (April 15, 2011). The comment period on the proposed rule is open until July 14, 2011.

[69] Kootenai Tribe of Idaho v. Dombeck, 142 F. Supp. 2d 1231(D. Idaho 2001), *rev'd,* 313 F. 3d 1094 (9th Cir. 2003).

[70] Idaho v. Dombeck, 142 F. Supp. 2d at 1248 (D. Idaho 2001), *rev'd,* 313 F. 3d 1094 (9th Cir. 2003).

[71] Kootenai Tribe of Idaho v. Dombeck, 142 F. Supp. 2d at 1244-1247.

[72] Kootenai Tribe of Idaho v. Veneman, CV01-10-N-EJL, 2001 WL 1141275 (D. Idaho May 10, 2001).

[73] Kootenai Tribe of Idaho v. Veneman, 313 F. 3d 1094, 1126 (9th Cir. 2003).

[74] Kootenai Tribe of Idaho v. Veneman, 313 F. 3d 1094, 1118-119 (9th Cir. 2003).

[75] *Id.* at 1120-121.

[76] Wyoming v. U.S. Department of Agriculture, 277 F. Supp. 2d 1197 (D. Wyo. 2003).

[77] *Id.* at 1231-232.

[78] *Id.* at 1203.

[79] *Id.* at 1232.

[80] *Id.* at 1239.

[81] Wyoming v. U.S. Department of Agriculture, 414 F.3d 1207 (10th Cir. 2005).

[82] California v. U.S. Department of Agriculture, C05-03508, 2006 WL 2711469 (N.D. Cal. September 20, 2006).

[83] Montana and Maine filed an amicus brief on behalf of plaintiffs; Alaska and Wyoming filed an amicus brief on behalf of defendants.

[84] California v. U.S. Department of Agriculture, 459 F. Supp. 2d 874 (N.D. Cal. 2006).

[85] *Id.* This order was issued in October, superseding the September decision. The injunction dates from September 20, 2006, however.

[86] California v. U.S. Department of Agriculture, 468 F. Supp. 2d 1140 (N.D. Cal. 2006).

[87] Wyoming v. U.S. Department of Agriculture, 277 F. Supp. 2d 1197 (D. Wyo. 2003).

[88] Wyoming v. U.S. Department of Agriculture, No. 01-CV-86 (D. Wyo. June 7, 2007).

[89] Wyoming v. U.S. Department of Agriculture, No. 01-CV-86, at 5 (D. Wyo. June 7, 2007).

[90] Wyoming v. U.S. Department of Agriculture, 570 F. Supp. 2d 1309 (D. Wyo. 2008).

[91] Wyoming v. U.S. Department of Agriculture, 570 F. Supp. 2d at 1352.

[92] California v. U.S. Department of Agriculture, 710 F. Supp. 2d 916 (N.D. Cal. 2008).

[93] Kootenai Tribe of Idaho v. Veneman, 313 F.3d 1094 (9th Cir. 2003) (holding that challenges to the NEPA review were not likely to succeed).

[94] California v. U.S. Department of Agriculture, 710 F. Supp. 2d 916, 924 (N.D. Cal. 2008).

[95] California v. U.S. Department of Agriculture, 575 F.3d 999 (9th Cir. 2009).

[96] Secretary's Memorandum 1042-154, Authority to Approve Road Construction and Timber Harvesting in Certain Lands Administered by the Forest Service (May 28, 2009), available online at http://fs.usda.gov/Internet/ FSE_DOCUMENTS/stelprdb5065998.pdf.

[97] Secretary's Memorandum 1042-155, Authority to Approve Road Construction and Timber Harvesting in Certain Lands Administered by the Forest Service (May 28, 2010), available online at http://www.fs.usda.gov/Internet/ FSE_DOCUMENTS/stelprdb5167211.pdf.

[98] Southeast Conference v. Vilsack, 684 F. Supp. 2d 185 (D.D.C. 2010).

[99] Organized Village of Kake v. U.S. Department of Agriculture, 1:09-cv-00023 JWS (D. Alaska March 4, 2011).

[100] Perhaps complicating the issue further, the regulation for the Idaho rule states that all of the states 9.3 million acres are "roadless," despite having classifications that allow development of roads, mining, forestry, or other development on much of the property. See 36 C.F.R. § 294.22(a). If a map also indicates all of these areas as roadless, the Senate bill would seem to nullify any such development.

INDEX

A

B

C

H

I

J

K

L

M

N

O

P

R

S

T

U

V

W

Y